Handbook of Genetics

ABOUT THE BOOK

Genetics is the study of heredity. Heredity is a biological process where a parent passes certain genes onto their children or offspring. Every child inherits genes from both of their biological parents and these genes in turn express specific traits. Some of these traits may be physical for example hair and eye color and skin color etc. On the other hand some genes may also carry the risk of certain diseases and disorders that may pass on from parents to their offspring. Genetics is the study of genes, genetic variation, and heredity in living organisms. It is generally considered a field of biology, but intersects frequently with many other life sciences and is strongly linked with the study of information systems.

The father of genetics is Gregor Mendel, a late 19th-century scientist and Augustinian friar. Mendel studied "trait inheritance", patterns in the way traits are handed down from parents to offspring. He observed that organisms (pea plants) inherit traits by way of discrete "units of inheritance". This term, still used today, is a somewhat ambiguous definition of what is referred to as a gene. Trait inheritance and molecular inheritance mechanisms of genes are still primary principles of genetics in the 21st century, but modern genetics has expanded beyond inheritance to studying the function and behavior of genes. Gene structure and function, variation, and distribution are studied within the context of the cell, the organism (e.g. dominance), and within the context of a population. Genetics has given rise to a number of subfields, including epigenetics and population genetics. Organisms studied within the broad field span the domain of life, including bacteria, plants, animals, and humans.

ABOUT THE AUTHOR

Dipnees Walkar, has worked widely in bother Management in Edinburgh-Scotland. He did his all Ph.D. from University of Aberdeen, Currently he is Professor in University of Dundee in geentics Department. His primary approach includes breaking the monoculture idea of agro biological systems through support of utilitarian biodiversity. Dipneed Walkar is writer of in excess of ninty nine logical papers and has likewise composed five books inculding Pest Management, Genetics in Microbiology and many more. His ebb and flow research investigates the utility of natural learning for more powerfull and reasonable plant insurance. He serves on the publication sheets of four journals and has distributed more than eleven referred papers and book parts and additionally co-altering the book Biological control.

Handbook of Genetics

DIPNEES WALKAR

WESTBURY PUBLISHING LTD.
ENGLAND (UNITED KINGDOM)

Handbook of Genetics
Edited by: Dipnees Walkar
ISBN: 978-1-913806-13-2 (Hardback)

© 2021 Westbury Publishing Ltd.

Published by **Westbury Publishing Ltd.**
Address: 6-7, St. John Street, Mansfield,
Nottinghamshire, England, NG18 1QH
United Kingdom
Email: - info@westburypublishing.com
Website: - www.westburypublishing.com

This book contains information obtained from authentic and highly regarded sources. All chapters are published with permission under the Creative Commons Attribution Share Alike License or equivalent. A Wide Variety of references are listed. Permissions and sources are indicated; for detailed attributions, please refer to the permission page. Reasonable efforts have been made to publish reliable data and information, but the authors, editors and publisher cannot assume any responsibility for the validity of the materials or the consequences of their use.

The publisher's policy is to use permanent paper from mills that operate a sustainable forestry policy. Furthermore, the publishers ensure that the text paper and cover boards used have met acceptable environmental accreditation standards.

Publisher Notice: - Presentations, Logos (the way they are written/ Presented), in this book are under the copyright of the publisher and hence, if copied/ resembled the copier will be prosecuted under the law.

British Library Cataloguing in Publication Data:
A catalogue record for this book is available from the British Library.

For more information regarding Westbury Publishing Ltd and its products, please visit the publisher's website- **www.westburypublishing.com**

Preface

Genetics is the study of heredity. Heredity is a biological process where a parent passes certain genes onto their children or offspring. Every child inherits genes from both of their biological parents and these genes in turn express specific traits. Some of these traits may be physical for example hair and eye color and skin color etc. On the other hand some genes may also carry the risk of certain diseases and disorders that may pass on from parents to their offspring. Genetics is the study of genes, genetic variation, and heredity in living organisms. It is generally considered a field of biology, but intersects frequently with many other life sciences and is strongly linked with the study of information systems.

The father of genetics is Gregor Mendel, a late 19th-century scientist and Augustinian friar. Mendel studied "trait inheritance", patterns in the way traits are handed down from parents to offspring. He observed that organisms (pea plants) inherit traits by way of discrete "units of inheritance". This term, still used today, is a somewhat ambiguous definition of what is referred to as a gene. Trait inheritance and molecular inheritance mechanisms of genes are still primary principles of genetics in the 21st century, but modern genetics has expanded beyond inheritance to studying the function and behavior of genes. Gene structure and function, variation, and distribution are studied within the context of the cell, the organism (e.g. dominance), and within the context of a population. Genetics has given rise to a number of subfields, including epigenetics and population genetics. Organisms studied within the broad field span the domain of life, including bacteria, plants, animals, and humans.

Genetic processes work in combination with an organism's environment and experiences to influence development and behavior, often referred to as nature versus nurture. The intracellular or extracellular environment of a cell or organism may switch gene transcription on or off. A classic example is two seeds of genetically identical corn, one placed in a temperate climate and one in an arid climate. While the average height of the two corn stalks may be genetically determined to be equal, the one in the arid climate only grows to half the height of the one in the temperate climate due to lack of water and nutrients in its environment.

—Editor

Contents

Preface (*v*)

1 **Basic Concepts of Genetics** 1
   - Introduction

2 **Molecular Genetics** 35
   - Introduction
   - Manipulating DNA
   - Manipulating Genes

3 **Molecular Biology** 68
   - Introduction
   - Structure of Protein
   - Protein Structure Determination

4 **Concepts of Cell and Molecular Biology** 85
   - Introduction
   - What is cell Biology?
   - Conclusion
   - Techniques of molecular biology

5 **Gene Expression and Regulation** 152
   (Recombinant of DNA and RNA Technology)
   - Introduction
   - Regulation of Gene Expression in Prokaryotes
   - Eukaryotic gene expression can be regulated at many stages

Index 191

1

Basic Concepts of Genetics

Genetics is the study of heredity. Heredity is a biological process where a parent passes certain genes onto their children or offspring. Every child inherits genes from both of their biological parents and these genes in turn express specific traits. Some of these traits may be physical for example hair and eye color and skin color etc. On the other hand some genes may also carry the risk of certain diseases and disorders that may pass on from parents to their offspring.

Genetics is the study of genes, genetic variation, and heredity in living organisms. It is generally considered a field of biology, but intersects frequently with many other life sciences and is strongly linked with the study of information systems.

The father of genetics is Gregor Mendel, a late 19th-century scientist and Augustinian friar. Mendel studied "trait inheritance", patterns in the way traits are handed down from parents to offspring. He observed that organisms (pea plants) inherit traits by way of discrete "units of inheritance". This term, still used today, is a somewhat ambiguous definition of what is referred to as a gene.

Trait inheritance and molecular inheritance mechanisms of genes are still primary principles of genetics in the 21st century, but modern genetics has expanded beyond inheritance to studying the function and behavior of genes. Gene structure and function, variation, and distribution are studied within the context of the cell, the organism (e.g. dominance), and within the context of a population. Genetics has given rise to a number of subfields, including epigenetics and population genetics. Organisms studied within the broad field span the domain of life, including bacteria, plants, animals, and humans.

Genetic processes work in combination with an organism's environment and experiences to influence development and behavior, often referred to as nature versus nurture. The intracellular or extracellular environment of a cell or organism may switch gene transcription on or off. A classic example is two seeds of genetically identical corn, one placed in a temperate climate and one in an arid climate. While the average height of the two corn stalks may be genetically determined to be equal, the one in the arid climate only grows to half the height of the one in the temperate climate due to lack of water and nutrients in its environment.

INTRODUCTION

Genetics is the study of genes—what they are, what they do, and how they work. Genes inside the nucleus of a cell are strung together in such a way that the sequence carries information: that information determines how living organisms inherit various features (phenotypic traits). For example, offspring produced by sexual reproduction usually look similar to each of their parents because they have inherited some of each of their parents' genes. Genetics identifies which features are inherited, and explains how these features pass from generation to generation. In addition to inheritance, genetics studies how genes are turned on and off to control what substances are made in a cell—gene expression; and how a cell divides—mitosis or meiosis.

Some phenotypic traits can be seen, such as eye color while others can only be detected, such as blood type or intelligence. Traits determined by genes can be modified by the animal's surroundings (environment): for example, the general design of a tiger's stripes is inherited, but the specific stripe pattern is determined by the tiger's surroundings. Another example is a person's height: it is determined by both genetics and nutrition.

Chromosomes are tiny packages which contain one DNA molecule and its associated proteins. Humans have 46 chromosomes (23 pairs). This number varies between species—for example, many primates have 24 pairs. Meiosis creates special cells, sperm in males and eggs in females, which only have 23 chromosomes. These two cells merge into one during the fertilization stage of sexual reproduction, creating a zygote. In a zygote, a nucleic acid double helix divides, with each single helix occupying one of the daughter cells, resulting in half the normal number of genes. By the time the zygote divides again, genetic recombination has created a new embryo with 23 pairs of chromosomes, half from each parent. Mating and resultant mate choice result in sexual selection. In normal cell division (mitosis) is possible when the double helix separates, and a complement of each separated half is made, resulting in two identical double helices in one cell, with each occupying one of the two new daughter cells created when the cell divides.

Basic Concepts of Genetics

Chromosomes all contain DNA made up of four nucleotides, abbreviated C (cytosine), G (guanine), A (adenine), or T (thymine), which line up in a particular sequence and make a long string. There are two strings of nucleotides coiled around one another in each chromosome: a double helix. C on one string is always opposite from G on the other string; A is always opposite T. There are about 3.2 billion nucleotide pairs on all the human chromosomes: this is the human genome. The order of the nucleotides carries genetic information, whose rules are defined by the genetic code, similar to how the order of letters on a page of text carries information. Three nucleotides in a row—a triplet—carry one unit of information: a codon.

The genetic code not only controls inheritance: it also controls gene expression, which occurs when a portion of the double helix is uncoiled, exposing a series of the nucleotides, which are within the interior of the DNA. This series of exposed triplets (codons) carries the information to allow machinery in the cell to "read" the codons on the exposed DNA, which results in the making of RNA molecules. RNA in turn makes either amino acids or microRNA, which are responsible for all of the structure and function of a living organism; i.e. they determine all the features of the cell and thus the entire individual. Closing the uncoiled segment turns off the gene.

Heritability means the information in a given gene is not always exactly the same in every individual in that species, so the same gene in different individuals does not give exactly the same instructions. Each unique form of a single gene is called an allele; different forms are collectively called polymorphisms. As an example, one allele for the gene for hair color and skin cell pigmentation could instruct the body to produce black pigment, producing black hair and pigmented skin; while a different allele of the same gene in a different individual could give garbled instructions that would result in a failure to produce any pigment, giving white hair and no pigmented skin: albinism. Mutations are random changes in genes creating new alleles, which in turn produce new traits, which could help, harm, or have no new effect on the individual's likelihood of survival; thus, mutations are the basis for evolution.

Actually Genetics is the branch of biology dealing with the phenomena of Heredity and the law governing it. biochemical genetics the study of the fundamental relationships between genes, protein, and metabolism. Thisinvolves the study of the cause of many specific heritable diseases. These include those resulting from the impropersynthesis of hemoglobins and protein, such as SICKLE CELL DISEASE and THALASSEMIA, both of which are hereditary anemias;some 200 inborn errors of metabolism, such as PHENYLKETONURIA and GALACTOSEMIA, in which lack or alteration of a specificenzyme prohibits proper metabolism of carbohydrates, proteins, or fats and thus produces pathologic symptoms; andgenetically determined variations in response to certain drugs, for example, isoniazid.

Clinical genetics the study of the causes and inheritance of genetic disorders. In addition to the diseases mentionedunder biochemical genetics, other aspects of clinical genetics include the study of chromosomal aberrations, such asthose that cause mental retardation and DOWN SYNDROME, and immunogenetics, or the genetic aspects of the IMMUNE RESPONSE and the transmission of genetic factors from generation to generation. Many pediatric hospital admissions involve genetic disorders. In obstetrics and neonatal medicine, prenatal diagnosis ofgenetic defects and improvement of pre and perinatal conditions are a major concern. In adults, such diseases asbreast cancer, coronary artery disease, hypertension, and diabetes mellitus have all been found to have predisposinggenetic components that are relevant to identification of risk factors and early diagnosis.

History of Genetics

For thousands of years, humans have acted as agents of genetic selection, by breeding offspring with desired traits. All our domesticated animals (dogs, horses, cattle) and food crops (wheat, corn) are the result. Yet for most of this time, humans had no idea how traits were inherited. Why? Offspring resemble parents (or don't) in bewilderingly complex ways. That is because individuals in nature contain many genes, and many different versions (alleles) of each gene.

Pre-Mendelian ideas on heredity

The most influential early theories of heredity were that of Hippocrates and Aristotle. Hippocrates' theory (possibly based on the teachings of Anaxagoras) was similar to Darwin's later ideas on pangenesis, involving heredity material that collects from throughout the body. Aristotle suggested instead that the (nonphysical) form-giving principle of an organism was transmitted through semen (which he considered to be a purified form of blood) and the mother's menstrual blood, which interacted in the womb to direct an organism's early development. For both Hippocrates and Aristotle—and nearly all Western scholars through to the late 19th century—the inheritance of acquired characters was a supposedly well-established fact that any adequate theory of heredity had to explain. At the same time, individual species were taken to have a fixed essence; such inherited changes were merely superficial.

In the Charaka Samhita of 300CE, ancient Indian medical writers saw the characteristics of the child as determined by four factors: 1) those from

Basic Concepts of Genetics

the mother's reproductive material, (2) those from the father's sperm, (3) those from the diet of the pregnant mother and (4) those accompanying the soul which enters into the foetus. Each of these four factors had four parts creating sixteen factors of which the karma of the parents and the soul determined which attributes predominated and thereby gave the child its characteristics.

In the 9th century CE, the Afro-Arab writer Al-Jahiz considered the effects of the environment on the likelihood of an animal to survive. In 1000 CE, the Arab physician, Abu al-Qasim al-Zahrawi (known as Albucasis in the West) was the first physician to describe clearly the hereditary nature of haemophilia in his Al-Tasrif. In 1140 CE, Judah HaLevi described dominant and recessive genetic traits in The Kuzari.

Plant systematics and hybridization

In the 18th century, with increased knowledge of plant and animal diversity and the accompanying increased focus on taxonomy, new ideas about heredity began to appear. Linnaeus and others (among them Joseph Gottlieb Kölreuter, Carl Friedrich von Gärtner, and Charles Naudin) conducted extensive experiments with hybridization, especially species hybrids. Species hybridizers described a wide variety of inheritance phenomena, include hybrid sterility and the high variability of back-crosses.

Plant breeders were also developing an array of stable varieties in many important plant species. In the early 19th century, Augustin Sageret established the concept of dominance, recognizing that when some plant varieties are crossed, certain characters (present in one parent) usually appear in the offspring; he also found that some ancestral characters found in neither parent may appear in offspring. However, plant breeders made little attempt to establish a theoretical foundation for their work or to share their knowledge with current work of physiology, although Gartons Agricultural Plant Breeders in England explained their system.

In 1866, Gregor Mendel discovered independent assortment of traits, dominant and recessive expression. Traits appear in pairs; separate independently in the gametes; recombine in pairs, in offspring. (Today we know Mendel only studied unlinked traits: on separate chromosomes, or so far apart that crossover frequency approached 50%). But Mendel's work was lost. Only in the past century did humans learn the fundamental mechanisms of heredity: How and why organisms resemble their parents; and how the inherited information functions to make organisms look and behave as they do. he observation that living things inherit traits from their parents has been used since prehistoric times to improve crop plants and animals through selective breeding.[6] The modern science of genetics, seeking to understand this process, began with the work of Gregor Mendel in the mid-19th century.

Prior to Mendel, Imre Festetics, a Hungarian noble, who lived in Koszeg before Mendel, was the first who used the word "genetics." He described several rules of genetic inheritance in his work The genetic law of the Nature (Die genetische Gesätze der Natur, 1819). His second law is the same as what Mendel published. In his third law, he developed the basic principles of mutation (he can be considered a forerunner of Hugo de Vries).

Other theories of inheritance preceded his work. A popular theory during Mendel's time was the concept of blending inheritance: the idea that individuals inherit a smooth blend of traits from their parents. Mendel's work provided examples where traits were definitely not blended after hybridization, showing that traits are produced by combinations of distinct genes rather than a continuous blend. Blending of traits in the progeny is now explained by the action of multiple genes with quantitative effects. Another theory that had some support at that time was the inheritance of acquired characteristics: the belief that individuals inherit traits strengthened by their parents. This theory (commonly associated with Jean-Baptiste Lamarck) is now known to be wrong—the experiences of individuals do not affect the genes they pass to their children, although evidence in the field of epigenetics has revived some aspects of Lamarck's theory. Other theories included the pangenesis of Charles Darwin (which had both acquired and inherited aspects) and Francis Galton's reformulation of pangenesis as both particulate and inherited.

Mendelian and classical genetics

Modern genetics started with Gregor Johann Mendel, a scientist and Augustinian friar who studied the nature of inheritance in plants. In his paper "Versuche über Pflanzenhybriden" ("Experiments on Plant Hybridization"), presented in 1865 to the Naturforschender Verein (Society for Research in Nature) in Brünn, Mendel traced the inheritance patterns of certain traits in pea plants and described them mathematically. Although this pattern of inheritance could only be observed for a few traits, Mendel's work suggested that heredity was particulate, not acquired, and that the inheritance patterns of many traits could be explained through simple rules and ratios.

The importance of Mendel's work did not gain wide understanding until the 1890s, after his death, when other scientists working on similar problems re-discovered his research. William Bateson, a proponent of Mendel's work, coined the word genetics in 1905 (the adjective genetic, derived from the

Basic Concepts of Genetics

Greek word genesis—???es??, "origin", predates the noun and was first used in a biological sense in 1860. Bateson both acted as a mentor and was aided significantly by the work of female scientists from Newnham College at Cambridge, specifically the work of Becky Saunders, Nora Darwin Barlow, and Muriel Wheldale Onslow. Bateson popularized the usage of the word genetics to describe the study of inheritance in his inaugural address to the Third International Conference on Plant Hybridization in London in 1906.

After the rediscovery of Mendel's work, scientists tried to determine which molecules in the cell were responsible for inheritance. In 1911, Thomas Hunt Morgan argued that genes are on chromosomes, based on observations of a sex-linked white eye mutation in fruit flies. In 1913, his student Alfred Sturtevant used the phenomenon of genetic linkage to show that genes are arranged linearly on the chromosome.

Some Other Important Development

1902 -- Walter Sutton and Theodore Boveri, using dyes synthesized by the German organic chemistry industry, observed that "colored bodies" in cells behaved in ways parallel to the hypothetical agents of heredity proposed by Mendel. These bodies were called chromosomes.

1905 -- Nettie Stevens observed in Tenebrio beetles that all pairs of homologous chromosomes are the same size, except for one pair which determines sex -- X, Y.

1909 -- Thomas H. Morgan correlates the X chromosome with sex-linked inheritance of the white eye trait in Drosophila -- a strain of flies discovered by an undergraduate lab assistant, cleaning out old bottles of flies in Morgan's lab. Morgan went on to make many important discoveries in fly genetics and linkage analysis that apply to all diploid organisms.

1941 -- Beadle and Tatum determined in Neurospora that each gene encodes one product (protein). (Later, we learned that RNA can be a product, not always transcribed to protein; for example, a ribosomal RNA.)

1944 -- Oswald Avery identified DNA as the genetic material. Pieces of DNA can transfer genes into bacteria cells, and transform them genetically.

1953 -- Rosalind Franklin and Maurice Wilkins showed that DNA is a double helix.

Thomas Watson and Frances Crick determined the structure of the base pairs which enable replication producing two identical daughter helices.

1961 -- Jacob and Monod figured out regulation of the lac operon.
1960's -- Barbara McClintockdiscovered transposable elements in corn; later found in bacteria and animals.
1970 -- Temin and Balitimore discovered reverse transcriptase in retroviruses; an enzyme later used to clone genes based on the RNA encoding the product.
1977 -- Maxam, Gilbert, Sanger, others -- developed methods to sequence DNA.
1981 -- The first transgenic mammals were made.
1987 -- Kary Mullis invented the polymerase chain reaction (PCR), using a thermostable enzyme from a thermophilic bacterium discovered by Thomas Brock at a geyser in Yellowstone. Mullis sold the process to a pharmaceutical company, and earned very little. Brock didn't earn a cent.
1995 -- The first bacterial genome sequence, Haemophilus influenzae, was completely determined.
1996 -- Ian Wilmut cloned the lamb Dolly from adult mammary gland tissue.

About Gragar John Mendal

Gregor Mendel was an Austrian monk who discovered the basic principles of heredity through experiments in his garden. Mendel's observations became the foundation of modern genetics and the study of heredity, and he is widely considered a pioneer in the field of genetics.

quotes

"My scientific studies have afforded me great gratification; and I am convinced that it will not be long before the whole world acknowledges the results of my work."

—Gregor Mendel

Synopsis

Gregor Mendel, known as the "father of modern genetics," was born in Austria in 1822. A monk, Mendel discovered the basic principles of heredity through experiments in his monastery's garden. His experiments showed that the inheritance of certain traits in pea plants follows particular patterns, subsequently becoming the foundation of modern genetics and leading to the study of heredity.

Early Life

Gregor Johann Mendel was born Johann Mendel on July 22, 1822, to Anton and Rosine Mendel, on his family's farm, in what was then Heinzendorf, Austria. He spent his early youth in that rural setting, until age 11, when a local schoolmaster who was impressed with his aptitude for learning recommended that he be sent to secondary school in Troppau to continue his education. The move was a financial strain on his family, and often a difficult experience for Mendel, but he excelled in his studies, and in 1840, he graduated from the school with honors.

Following his graduation, Mendel enrolled in a two-year program at the Philosophical Institute of the University of Olmütz. There, he again distinguished himself academically, particularly in the subjects of physics and math, and tutored in his spare time to make ends meet. Despite suffering from deep bouts of depression that, more than once, caused him to temporarily abandon his studies, Mendel graduated from the program in 1843.

That same year, against the wishes of his father, who expected him to take over the family farm, Mendel began studying to be a monk: He joined the Augustinian order at the St. Thomas Monastery in Brno, and was given the name Gregor. At that time, the monastery was a cultural center for the region, and Mendel was immediately exposed to the research and teaching of its members, and also gained access to the monastery's extensive library and experimental facilities.

In breeding experiments between 1856 and 1865, Gregor Mendel first traced inheritance patterns of certain traits in pea plants and showed that they obeyed simple statistical rules with some traits being dominant and others being recessive. These patterns of Mendelian inheritance demonstrated that application of statistics to inheritance could be highly useful; they also contradicted 19th century theories of blending inheritance as the traits remained discrete through multiple generation of hybridization. Since that time many more complex forms of inheritance have been demonstrated.

From his statistical analysis Mendel defined a concept that he described as a character (which in his mind holds also for "determinant of that character"). In only one sentence of his historical paper he used the term "factors" to designate the "material creating" the character: " So far as experience goes, we find it in every case confirmed that constant progeny can only be formed when the egg cells and the fertilizing pollen are of like character, so that both are provided with the material for creating quite similar individuals, as is the case with the normal fertilization of pure species. We must therefore regard it as certain that exactly similar factors must be at work also in the production of the constant forms in the hybrid plants."(Mendel, 1866).

Mendel's work was published in 1866 as "Versuche über Pflanzen-Hybriden" (Experiments on Plant Hybridization) in the Verhandlungen des Naturforschenden Vereins zu Brünn (Proceedings of the Natural History Society of Brünn), following two lectures he gave on the work in early 1866.

Later Life and Legacy

In 1868, Mendel was elected abbot of the school where he had been teaching for the previous 14 years, and both his resulting administrative duties and his gradually failing eyesight kept him from continuing any extensive scientific work. He traveled little during this time, and was further isolated from his contemporaries as the result of his public opposition to an 1874 taxation law that increased the tax on the monasteries to cover Church expenses.

Gregor Mendel died on January 6, 1884, at the age of 61. He was laid to rest in the monastery's burial plot and his funeral was well attended. His work, however, was still largely unknown.

It was not until decades later, when Mendel's research informed the work of several noted geneticists, botanists and biologists conducting research on heredity, that its significance was more fully appreciated, and his studies began to be referred to as Mendel's Laws. Hugo de Vries, Carl Correns and Erich von Tschermak-Seysenegg each independently duplicated Mendel's experiments and results in 1900, finding out after the fact, allegedly, that both the data and the general theory had been published in 1866 by Mendel. Questions arose about the validity of the claims that the trio of botanists were not aware of Mendel's previous results, but they soon did credit Mendel with priority. Even then, however, his work was often marginalized by Darwinians, who claimed that his findings were irrelevant to a theory of evolution. As genetic theory continued to develop, the relevance of Mendel's work fell in and out of favor, but his research and theories are considered fundamental to any understanding of the field, and he is thus considered the "father of modern genetics."

Experiments and Theories

Around 1854, Mendel began to research the transmission of hereditary traits in plant hybrids. At the time of Mendel's studies, it was a generally accepted fact that the hereditary traits of the offspring of any species were merely the diluted blending of whatever traits were present in the "parents." It was also commonly accepted that, over generations, a hybrid would revert to its original form, the implication of which suggested that a hybrid could not create new forms. However, the results of such studies were often skewed by the relatively short period of time during which the experiments were

Basic Concepts of Genetics

conducted, whereas Mendel's research continued over as many as eight years (between 1856 and 1863), and involved tens of thousands of individual plants.

Mendel chose to use peas for his experiments due to their many distinct varieties, and because offspring could be quickly and easily produced. He cross-fertilized pea plants that had clearly opposite characteristics—tall with short, smooth with wrinkled, those containing green seeds with those containing yellow seeds, etc.—and, after analyzing his results, reached two of his most important conclusions: the Law of Segregation, which established that there are dominant and recessive traits passed on randomly from parents to offspring (and provided an alternative to blending inheritance, the dominant theory of the time), and the Law of Independent Assortment, which established that traits were passed on independently of other traits from parent to offspring. He also proposed that this heredity followed basic statistical laws. Though Mendel's experiments had been conducted with pea plants, he put forth the theory that all living things had such traits.

In 1865, Mendel delivered two lectures on his findings to the Natural Science Society in Brno, who published the results of his studies in their journal the following year, under the title Experiments on Plant Hybrids. Mendel did little to promote his work, however, and the few references to his work from that time period indicated that much of it had been misunderstood. It was generally thought that Mendel had shown only what was already commonly known at the time—that hybrids eventually revert to their original form. The importance of variability and its evolutionary implications were largely overlooked. Furthermore, Mendel's findings were not viewed as being generally applicable, even by Mendel himself, who surmised that they only applied to certain species or types of traits. Of course, his system eventually proved to be of general application and is one of the foundational principles of biology

Post-Mendel, pre-re-discovery

Mendel's work was published in a relatively obscure scientific journal, and it was not given any attention in the scientific community. Instead, discussions about modes of heredity were galvanized by Darwin's theory of evolution by natural selection, in which mechanisms of non-Lamarckian heredity seemed to be required. Darwin's own theory of heredity, pangenesis, did not meet with any large degree of acceptance. A more mathematical version of pangenesis, one which dropped much of Darwin's Lamarckian holdovers, was developed as the "biometrical" school of heredity by Darwin's cousin, Francis Galton. Under Galton and his successor Karl Pearson, the biometrical school attempted to build statistical models for heredity and evolution, with some limited but real success, though the exact methods of heredity were unknown and largely unquestioned.

In 1883 August Weismann conducted experiments involving breeding mice whose tails had been surgically removed. His results — that surgically removing a mouse's tail had no effect on the tail of its offspring — challenged the theories of pangenesis and Lamarckism, which held that changes to an organism during its lifetime could be inherited by its descendants. Weismann proposed the germ plasm theory of inheritance, which held that hereditary information was carried only in sperm and egg cells.

Re-discovery of Mendel

Hugo de Vries wondered what the nature of germ plasm might be, and in particular he wondered whether or not germ plasm was mixed like paint or whether the information was carried in discrete packets that remained unbroken. In the 1890s he was conducting breeding experiments with a variety of plant species and in 1897 he published a paper on his results that stated that each inherited trait was governed by two discrete particles of information, one from each parent, and that these particles were passed along intact to the next generation. In 1900 he was preparing another paper on his further results when he was shown a copy of Mendel's 1866 paper by a friend who thought it might be relevant to de Vries's work. He went ahead and published his 1900 paper without mentioning Mendel's priority. Later that same year another botanist, Carl Correns, who had been conducting hybridization experiments with maize and peas, was searching the literature for related experiments prior to publishing his own results when he came across Mendel's paper, which had results similar to his own. Correns accused de Vries of appropriating terminology from Mendel's paper without crediting him or recognizing his priority. At the same time another botanist, Erich von Tschermak was experimenting with pea breeding and producing results like Mendel's. He too discovered Mendel's paper while searching the literature for relevant work. In a subsequent paper de Vries praised Mendel and acknowledged that he had only extended his earlier work.

Genes and inheritance

The information gained from the Human Genome Project and related genetic research will undoubtedly create significant changes in healthcare practice. It is becoming increasingly clear that nurses in all areas of clinical practice will require a fundamental understanding of basic genetics. This article provides the oncology nurse with an overview of basic genetic concepts, including inheritance patterns of single gene conditions, pedigree construction, chromosome aberrations, and the multifactorial basis underlying the common

Basic Concepts of Genetics

diseases of adulthood. Normal gene structure and function are be introduced and the biochemistry of genetic errors is described

Genes are pieces of DNA that contain information for synthesis of ribonucleic acids (RNAs) or polypeptides. Genes are inherited as units, with two parents dividing out copies of their genes to their offspring. This process can be compared with mixing two hands of cards, shuffling them, and then dealing them out again. Humans have two copies of each of their genes, and make copies that are found in eggs or sperm—but they only include one copy of each type of gene. An egg and sperm join to form a complete set of genes. The eventually resulting offspring has the same number of genes as their parents, but for any gene one of their two copies comes from their father, and one from their mother.

The effects of this mixing depend on the types (the alleles) of the gene. If the father has two copies of an allele for red hair, and the mother has two copies for brown hair, all their children get the two alleles that give different instructions, one for red hair and one for brown. The hair color of these children depends on how these alleles work together. If one allele dominates the instructions from another, it is called the dominant allele, and the allele that is overridden is called the recessive allele. In the case of a daughter with alleles for both red and brown hair, brown is dominant and she ends up with brown hair.

Although the red color allele is still there in this brown-haired girl, it doesn't show. This is a difference between what you see on the surface (the traits of an organism, called its phenotype) and the genes within the organism (its genotype). In this example you can call the allele for brown "B" and the allele for red "b". (It is normal to write dominant alleles with capital letters and recessive ones with lower-case letters.) The brown hair daughter has the "brown hair phenotype" but her genotype is Bb, with one copy of the B allele, and one of the b allele.

Now imagine that this woman grows up and has children with a brown-haired man who also has a Bb genotype. Her eggs will be a mixture of two types, one sort containing the B allele, and one sort the b allele. Similarly, her partner will produce a mix of two types of sperm containing one or the other of these two alleles. When the transmitted genes are joined up in their offspring, these children have a chance of getting either brown or red hair, since they could get a genotype of BB = brown hair, Bb = brown hair or bb = red hair. In this generation, there is therefore a chance of the recessive allele showing itself in the phenotype of the children—some of them may have red hair like their grandfather.

Many traits are inherited in a more complicated way than the example above. This can happen when there are several genes involved, each contributing a small part to the end result. Tall people tend to have tall children because their children get a package of many alleles that each contribute a bit to how much they grow. However, there are not clear groups of "short people" and "tall people", like there are groups of people with brown or red hair. This is because of the large number of genes involved; this makes the trait very variable and people are of many different heights. Despite a common misconception, the green/blue eye traits are also inherited in this complex inheritance model. Inheritance can also be complicated when the trait depends on interaction between genetics and environment. For example, malnutrition does not change traits like eye color, but can stunt growth.

What is a gene?

Genes are special segments of DNA letters that, when read correctly by the body's proteins, can provide a specific and important instruction for the body to function properly. Researchers estimate that there are about 22,000 genes contained in the genome. Although genes are very important, they make up only a small percentage of all of the DNA in the genome. Each gene has a specific location on one of our 23 chromosomes and is inherited, or passed down, from generation to generation as a unit. We have two copies of each chromosome and, thus, two copies of each gene. We inherit one copy from each of our parents and, in turn, pass on one of our two copies to each of our children. Each gene contains a specific set of instructions for the body. Some genes contain multiple sets of instructions. Usually these instructions make a protein. There are many different types of proteins in our bodies which can perform multiple important tasks. For example, proteins form the basis of our organ tissues, bones, and nervous system. They also guide how we digest food and medications.

Genes in the cell

The genetic information lies within the cell nucleus of each living cell in the body. The information can be considered to be retained in a book for example. Part of this book with the genetic information comes from the father while the other part comes from the mother.

Chromosomes

The genes lie within the chromosomes. Humans have 23 pairs of these small thread-like structures in the nucleus of their cells. 23 or half of the total 46 comes from the mother while the other 23 comes from the father. The chromosomes contain genes just like pages of a book. Some chromosomes may carry thousands of important genes while some may carry only a few. The chromosomes, and therefore the genes, are made up of the chemical substance called DNA (DeoxyriboNucleic Acid). The chromosomes are very long thin strands of DNA, coiled up tightly.

At one point along their length, each chromosome has a constriction, called the centromere. The centromere divides the chromosomes into two 'arms': a long arm and a short arm. Chromosomes are numbered from 1 to 22 and these are common for both sexes and called autosomes. There are also two chromosomes that have been given the letters X and Y and termed sex chromosomes. The X chromosome is much larger than the Y chromosome.

Chemical bases

The genes are further made up of unique codes of chemical bases comprising of A, T, C and G (Adenine, Thymine, Cytosine and Guanine). These chemical bases make up combinations with permutations and combinations. These are akin to the words on a page.

These chemical bases are part of the DNA. The words when stringed together act as the blueprints that tells the cells of the body when and how to grow, mature and perform various functions. With age the genes may be affected and may develop faults and damages due to environmental and endogenous toxins.

What is an inherited disease?

Some diseases are hereditary and run in families; others, such as infectious diseases, are caused by the environment. Other diseases come from a combination of genes and the environment. Genetic disorders are diseases that are caused by a single allele of a gene and are inherited in families. These include Huntington's disease, Cystic fibrosis or Duchenne muscular dystrophy. Cystic fibrosis, for example, is caused by mutations in a single gene called CFTR and is inherited as a recessive trait.

Other diseases are influenced by genetics, but the genes a person gets from their parents only change their risk of getting a disease. Most of these diseases are inherited in a complex way, with either multiple genes involved, or coming from both genes and the environment. As an example, the risk of breast cancer is 50 times higher in the families most at risk, compared to the families least at risk. This variation is probably due to a large number of alleles, each changing the risk a little bit. Several of the genes have been identified, such as BRCA1 and BRCA2, but not all of them. However, although some of the risk is genetic, the risk of this cancer is also increased by being overweight, drinking a lot of alcohol and not exercising. A woman's risk of breast cancer therefore comes from a large number of alleles interacting with her environment, so it is very hard to predict.

Each cell in the body contains 23 pairs of chromosomes. One chromosome from each pair is inherited from your mother and one is inherited from your father. The chromosomes contain the genes you inherit from your parents. ... whether the gene for that condition is dominant or recessive.

The chromosomes contain the genes you inherit from your parents. For example, for the gene that determines eye colour, you may inherit a brown-eye gene from your mother and a blue-eye gene from your father. In this instance, you'll end up with brown eyes because brown is the dominant gene. The different forms of genes for eye colour are caused by changes (mutations) in the DNA code. The same is true for medical conditions. There may be a faulty version of a gene that results in a medical condition, and a normal version that may not cause health problems. Whether your child ends up with a medical condition will depend on several factors, including:
- what genes they inherit
- whether the gene for that condition is dominant or recessive
- their environment, including any preventative treatment they may receive

Passing on mutations

Some medical conditions are directly caused by a mutation in a single gene that may have been passed on to a child by his or her parents.

Depending on the specific condition concerned, these genetic conditions can be inherited in three main ways, outlined below.

Autosomal recessive inheritance

Some conditions can only be inherited in an autosomal recessive pattern. This means the condition can only be passed on to a child if both parents have a copy of the faulty gene – both are "carriers" of the condition.

If the child only inherits one copy of the faulty gene, they'll be a carrier of the condition but won't have the condition.

If a mother and a father both carry the faulty gene, there's a one in four (25%) chance of each child they have inheriting the genetic condition, and a one in two chance (50%) of their child being a carrier.

Examples of genetic conditions inherited in this way include:
- cystic fibrosis – a condition in which the lungs and digestive system become clogged with thick, sticky mucus
- sickle cell anaemia – a condition where red blood cells, which carry oxygen around the body, develop abnormally

- thalassaemia – a group of conditions where the part of the blood known as haemoglobin is abnormal, which means affected red blood cells are unable to function normally
- Tay-Sachs disease – a condition that causes progressive damage to the nervous system

Autosomal dominant inheritance

Some conditions are inherited in an autosomal dominant pattern. In this case, only one parent needs to carry the mutation for the condition to be passed on to the child.

If one parent has the mutation, there's a one in two (50%) chance it will be passed on to each child the couple has.

Examples of genetic conditions inherited in this way include:

- type 1 neurofibromatosis – a condition that causes tumours to grow along the nerves
- tuberous sclerosis – a condition that causes mainly non-cancerous (benign) tumours to develop in different parts of the body
- Huntingdon's disease – a condition where certain brain cells become increasingly damaged over time
- autosomal dominant polycystic kidney disease (ADPKD) – a condition that causes small, fluid-filled sacs called cysts to develop in the kidneys

X-linked inheritance

Some conditions are caused by a mutation on the X chromosome (one of the sex chromosomes). These are usually inherited in a recessive pattern, but in a slightly different way from the autosomal recessive pattern described above.

X-linked recessive conditions often don't affect females to a significant degree because they have two X chromosomes, one of which will almost certainly be normal and can usually compensate for the mutated chromosome. However, females who inherit the mutation will become carriers.

Males can't inherit X-linked mutations from their fathers because they receive a Y chromosome from them. A male will develop the condition if he inherits the mutation from his mother. This is because he doesn't have the normal X chromosome to compensate.

When a mother is a carrier of an X-linked mutation, each daughter they have has a one in two (50%) chance of becoming a carrier, and each son they have has a one in two (50%) chance of inheriting the condition.

When a father has an X-linked condition, his sons won't be affected because he'll pass on a Y chromosome to them. However, any daughters he has will become carriers of the mutation.

Examples of genetic conditions inherited in this way include:
- Duchenne muscular dystrophy – a condition that causes the muscles to gradually weaken, resulting in an increasing level of disability
- haemophilia – a condition that affects the blood's ability to clot
- fragile X syndrome – a condition that usually causes certain facial and bodily characteristics, such as a long face, large ears and flexible joints

New mutations

Although genetic conditions are often inherited, this isn't always the case.

Some genetic mutations can occur for the first time when a sperm or egg is made, when a sperm fertilises an egg, or when cells are dividing after fertilisation. This is known as a de novo, or sporadic, mutation.

Someone with a new mutation won't have a family history of a condition, but they may be at risk of passing the mutation on to their children.

They may also have, or be at risk of developing, a form of the condition themselves.

Examples of conditions that are often caused by a de novo mutation include some types of muscular dystrophy, haemophilia and type 1 neurofibromatosis.

Chromosomal conditions

Some conditions aren't caused by a mutation on a specific gene, but by an abnormality in a person's chromosomes, such as having too many or too few chromosomes, rather than the normal 23 pairs.

Examples of conditions caused by chromosomal abnormalities include:
- Down's syndrome – caused by having an extra copy of chromosome 21
- Edwards' syndrome – caused by having an extra copy of chromosome 18
- Patau's syndrome – caused by having an extra copy of chromosome 13
- Turner syndrome – a disorder that only affects females and is caused by a missing or abnormal X chromosome
- Klinefelter's syndrome – a disorder that only affects males and is caused by an extra X chromosome

While these are genetic conditions, they're generally not inherited. They usually occur randomly as a result of a problem before, during or soon after the fertilisation of an egg by a sperm.

Influence of the environment

Very few health conditions are only caused by genes – most are caused by the combination of genes and environmental factors. Environmental factors include lifestyle factors, such as diet and exercise.

Around a dozen or so genes determine most human characteristics, such as height and the likelihood of developing common conditions.

Genes can have many variants, and studies of the whole genome – the whole set of genes – in large numbers of people are showing these variants may increase or decrease a person's chance of having certain conditions.

Each variant may only increase or decrease the chance of a condition very slightly, but this can add up across several genes.

In most people, the gene variants balance out to give an average risk for most conditions. But in some cases the risk is significantly above or below the average.

It's thought it may be possible to reduce the risk by changing environmental and lifestyle factors.

For example, coronary heart disease – when the heart's blood supply is blocked or interrupted – can run in families, but a poor diet, smoking and a lack of exercise can also increase your risk of developing the condition.

Research suggests that in the future, individuals will be able to find out what conditions they're most likely to develop.

It may then be possible to significantly reduce the chances of developing these conditions by making appropriate lifestyle and environmental changes.

Although genetic factors play a part in nearly all health conditions and characteristics, there are some conditions in which the genetic changes are almost exclusively responsible for causing the condition. These are called genetic disorders, or inherited diseases.

Since genes are passed from parent to child, any changes to the DNA within a gene are also passed. DNA changes may also happen spontaneously, showing up for the first time within the child of unaffected parents. This is referred to as a new mutation, where the word mutation means change. Sometimes this change can cause mistakes in the protein instructions, leading to production of a protein that doesn't work properly or cannot be made at all. When one protein is missing or not working as it should, it can cause a genetic disorder.

The genetics of each disorder are unique. In some cases, all the mistakes in a particular gene cause one specific genetic disorder. In other cases, different changes within the same gene can lead to different health or developmental problems or even to different genetic disorders. Sometimes changes in several similar genes may all lead to the same genetic disorder.

In other words a genetic disease is any disease that is caused by an abnormality in an individual's genome, the person's entire genetic makeup. The abnormality can range from minuscule to major -- from a discrete mutation in a single base in the DNA of a single gene to a gross chromosome abnormality involving the addition or subtraction of an entire chromosome or set of chromosomes. Some genetic disorders are inherited from the parents, while other genetic diseases are caused by acquired changes or mutations in a preexisting gene or group of genes. Mutations can occur either randomly or due to some environmental exposure.

Why does an inherited disease appear only some of the time?

Genetic disorders are typically inherited (passed down) in either a dominant or recessive manner. We each have two copies of every gene on our 22 numbered chromosomes. In addition, females have two copies of all the genes on the X chromosome, whereas males have one copy of the X chromosome genes and one copy of the Y chromosome genes.

When a disorder is dominant, the disease can occur when there are DNA mistakes in only one of the two gene copies. This means that if a parent has the DNA change, there is a 50-50 chance that it will be passed on to each child.

When a disorder is recessive, there must be mistakes in both copies of the gene for the disorder to occur. This means that both parents must carry at least one copy of the specific gene change in order to produce an affected child. If both parents have one changed copy, there is a 1 in 4, or 25% chance, that a child may inherit both changed copies at the same time, causing the disorder in the child. Parents who have only one changed gene copy usually do not display any symptoms of the disorder and may not even know they carry a gene change. Researchers estimate that we each have ("carry") 6-10 recessive gene changes. Certain recessive gene changes may be more common in different population groups. For example, sickle cell gene changes are found more often in individuals with West African ancestry and cystic fibrosis gene changes are more common in individuals with North European ancestry.

In addition to the inheritance pattern, some genetic disorders may be inconsistent when it comes to whether a person develops symptoms and their severity. Penetrance refers to whether the person who has the causative gene changes actually develops any symptoms of the disorder. Expressivity refers to the symptoms that may develop and their severity.

Basic Concepts of Genetics 21

How genes work

Each gene has a special job to do. The DNA in a gene spells out specific instructions—much like in a cookbook recipe — for making proteins (say: PRO-teens) in the cell. Proteins are the building blocks for everything in your body. Bones and teeth, hair and earlobes, muscles and blood, are all made up of proteins.

People have known for many years that living things inherit traits from their parents. That common-sense observation led to agriculture, the purposeful breeding and cultivation of animals and plants for desirable characteristics. Firming up the details took quite some time, though. Researchers did not understand exactly how traits were passed to the next generation until the middle of the 20th century.

Now it is clear that genes are what carry our traits through generations and that genes are made of deoxyribonucleic acid (DNA). But genes themselves don't do the actual work. Rather, they serve as instruction books for making functional molecules such as ribonucleic acid (RNA) and proteins, which perform the chemical reactions in our bodies.

Proteins do many other things, too. They provide the body's main building materials, forming the cell's architecture and structural components. But one thing proteins can't do is make copies of themselves. When a cell needs more proteins, it uses the manufacturing instructions coded in DNA.

The DNA code of a gene—the sequence of its individual DNA building blocks, labeled A (adenine), T (thymine), C (cytosine) and G (guanine) and collectively called nucleotides— spells out the exact order of a protein's building blocks, amino acids.

Occasionally, there is a kind of typographical error in a gene's DNA sequence. This mistake— which can be a change, gap or duplication—is called a mutation.

A mutation can cause a gene to encode a protein that works incorrectly or that doesn't work at all. Sometimes, the error means that no protein is made.

But not all DNA changes are harmful. Some mutations have no effect, and others produce new versions of proteins that may give a survival advantage to the organisms that have them. Over time, mutations supply the raw material from which new life forms evolve.

Genes make proteins

The function of genes is to provide the information needed to make molecules called proteins in cells. Cells are the smallest independent parts of organisms: the human body contains about 100 trillion cells, while very small organisms like bacteria are just one single cell. A cell is like a miniature and very complex factory that can make all the parts needed to produce a copy of itself, which happens when cells divide. There is a simple division of labor in cells—genes give instructions and proteins carry out these instructions, tasks like building a new copy of a cell, or repairing damage. Each type of protein is a specialist that only does one job, so if a cell needs to do something new, it must make a new protein to do this job. Similarly, if a cell needs to do something faster or slower than before, it makes more or less of the protein responsible. Genes tell cells what to do by telling them which proteins to make and in what amounts. Proteins are made of a chain of 20 different types of amino acid molecules. This chain folds up into a compact shape, rather like an untidy ball of string. The shape of the protein is determined by the sequence of amino acids along its chain and it is this shape that, in turn, determines what the protein does. For example, some proteins have parts of their surface that perfectly match the shape of another molecule, allowing the protein to bind to this molecule very tightly. Other proteins are enzymes, which are like tiny machines that alter other molecules.

The information in DNA is held in the sequence of the repeating units along the DNA chain. These units are four types of nucleotides (A,T,G and C) and the sequence of nucleotides stores information in an alphabet called the genetic code. When a gene is read by a cell the DNA sequence is copied into a very similar molecule called RNA (this process is called transcription). Transcription is controlled by other DNA sequences (such as promoters), which show a cell where genes are, and control how often they are copied. The RNA copy made from a gene is then fed through a structure called a ribosome, which translates the sequence of nucleotides in the RNA into the correct sequence of amino acids and joins these amino acids together to make a complete protein chain. The new protein then folds up into its active form. The process of moving information from the language of RNA into the language of amino acids is called translation.

If the sequence of the nucleotides in a gene changes, the sequence of the amino acids in the protein it produces may also change—if part of a gene is deleted, the protein produced is shorter and may not work any more. This is the reason why different alleles of a gene can have different effects in an organism. As an example, hair color depends on how much of a dark substance called melanin is put into the hair as it grows. If a person has a normal set of the genes involved in making melanin, they make all the proteins needed and they grow dark hair. However, if the alleles for a particular protein have different sequences and produce proteins that can't do their jobs, no melanin is produced and the person has white skin and hair (albinism).

Emergence of molecular genetics

After the rediscovery of Mendel's work there was a feud between William Bateson and Pearson over the hereditary mechanism, solved by Ronald Fisher in his work "The Correlation Between Relatives on the Supposition of Mendelian Inheritance".

In 1910, Thomas Hunt Morgan showed that genes reside on specific chromosomes. He later showed that genes occupy specific locations on the chromosome. With this knowledge, Morgan and his students began the first chromosomal map of the fruit fly Drosophila. In 1928, Frederick Griffith showed that genes could be transferred. In what is now known as Griffith's experiment, injections into a mouse of a deadly strain of bacteria that had been heat-killed transferred genetic information to a safe strain of the same bacteria, killing the mouse.

A series of subsequent discoveries led to the realization decades later that the genetic material is made of DNA (deoxyribonucleic acid). In 1941, George Wells Beadle and Edward Lawrie Tatum showed that mutations in genes caused errors in specific steps in metabolic pathways. This showed that specific genes code for specific proteins, leading to the "one gene, one enzyme" hypothesis. Oswald Avery, Colin Munro MacLeod, and Maclyn McCarty showed in 1944 that DNA holds the gene's information. In 1952, Rosalind Franklin and Raymond Gosling produced a strikingly clear x-ray diffraction pattern indicating a helical form, and in 1953, James D. Watson and Francis Crick demonstrated the molecular structure of DNA. Together, these discoveries established the central dogma of molecular biology, which states that proteins are translated from RNA which is transcribed by DNA. This dogma has since been shown to have exceptions, such as reverse transcription in retroviruses.

In 1972, Walter Fiers and his team at the University of Ghent were the first to determine the sequence of a gene: the gene for bacteriophage MS2 coat protein. Richard J. Roberts and Phillip Sharp discovered in 1977 that genes can be split into segments. This led to the idea that one gene can make several proteins. The successful sequencing of many organisms' genomes has complicated the molecular definition of genes. In particular, genes do not seem to sit side by side on DNA like discrete beads. Instead, regions of the DNA producing distinct proteins may overlap, so that the idea emerges that "genes are one long continuum". It was first hypothesized in 1986 by Walter Gilbert that neither DNA nor protein would be required in such a primitive system as that of a very early stage of the earth if RNA could perform as simply a catalyst and genetic information storage processor.

The modern study of genetics at the level of DNA is known as molecular genetics and the synthesis of molecular genetics with traditional Darwinian evolution is known as the modern evolutionary synthesis.

DNA

Up until the 1950s, scientists knew a good deal about heredity, but they didn't have a clue what DNA looked like. In order to learn more about DNA and its structure, some scientists experimented with using X rays as a form of molecular photography.

Rosalind Franklin, a physical chemist working with Maurice Wilkins at King's College in London, was among the first to use this method to analyze genetic material. Her experiments produced what were referred to at the time as "the most beautiful X-ray photographs of any substance ever taken."

Other scientists, including zoologist James Watson and physicist Francis Crick, both working at Cambridge University in the United Kingdom, were trying to determine the shape of DNA too. Ultimately, this line of research revealed one of the most profound scientific discoveries of the 20th century: that DNA exists as a double helix.

The 1962 Nobel Prize in physiology or medicine was awarded to Watson, Crick and Wilkins for this work. Although Franklin did not earn a share of the prize due to her untimely death at age 38, she is widely recognized as having played a significant role in the discovery.

Long strings of nucleotides form genes, and groups of genes are packaged tightly into structures called chromosomes. Every cell in your body except for eggs, sperm and red blood cells contains a full set of chromosomes in its nucleus.

If the chromosomes in one of your cells were uncoiled and placed end to end, the DNA would be about 6 feet long. If all the DNA in your body were connected in this way, it would stretch approximately 67 billion miles! That's nearly 150,000 round trips to the Moon.

Genes are copied

Genes are copied each time a cell divides into two new cells. The process that copies DNA is called DNA replication. It is through a similar process that a child inherits genes from its parents, when a copy from the mother is mixed with a copy from the father.

DNA can be copied very easily and accurately because each piece of DNA can direct the creation of a new copy of its information. This is because DNA is made of two strands that pair together like the two sides of a zipper. The nucleotides are in the center, like the teeth in the zipper, and pair up to hold

Basic Concepts of Genetics

the two strands together. Importantly, the four different sorts of nucleotides are different shapes, so for the strands to close up properly, an A nucleotide must go opposite a T nucleotide, and a G opposite a C. This exact pairing is called base pairing.

When DNA is copied, the two strands of the old DNA are pulled apart by enzymes; then they pair up with new nucleotides and then close. This produces two new pieces of DNA, each containing one strand from the old DNA and one newly made strand. This process is not predictably perfect as proteins attach to a nucleotide while they are building and cause a change in the sequence of that gene. These changes in DNA sequence are called mutations. Mutations produce new alleles of genes. Sometimes these changes stop the functioning of that gene or make it serve another advantageous function, such as the melanin genes discussed above. These mutations and their effects on the traits of organisms are one of the causes of of evolution.

Genetic Engineering

Since traits come from the genes in a cell, putting a new piece of DNA into a cell can produce a new trait. This is how genetic engineering works. For example, rice can be given genes from a maize and a soil bacteria so the rice produces beta-carotene, which the body converts to Vitamin A. This can help children suffering from Vitamin A deficiency. Another gene being put into some crops comes from the bacterium Bacillus thuringiensis; the gene makes a protein that is an insecticide. The insecticide kills insects that eat the plants, but is harmless to people. In these plants, the new genes are put into the plant before it is grown, so the genes are in every part of the plant, including its seeds. The plant's offspring inherit the new genes, which has led to concern about the spread of new traits into wild plants.

The kind of technology used in genetic engineering is also being developed to treat people with genetic disorders in an experimental medical technique called gene therapy. However, here the new gene is put in after the person has grown up and become ill, so any new gene is not inherited by their children. Gene therapy works by trying to replace the allele that causes the disease with an allele that works properly.

Genetic Engineering is also referred as genetic modification. It is a process of manually adding new DNA to a living organism through artificial methods.

Genetic Engineering is a method of physically removing a gene from one organism and inserting it to another and giving it the ability to express the qualities given by that gene.

ome examples of genetic engineering are Faster-growing trees, Bigger, longer-lasting tomatoes, Glow in the dark cats, Golden rice, Plants that fight pollution, banana vaccine, etc.

"Genetic Engineering is that field which is related to genes & DNA. Genetic engineering is used by scientists to improve or modify the traits of an individual organism".

An organism which is generated by applying genetic engineering is called as genetically modified organism (GMO). The first GMO were Bacteria generated in 1973 and GM mice in 1974. The techniques of genetic engineering have been applied in various fields such as research, agriculture, industrial biotechnology, and medicine. Genetic engineering focuses on biochemistry, cell biology, molecular biology, evolutionary biology, and medical genetics.

The term "genetic engineering" was firstly used by Jack Williamson in Dragons Island a science fiction novel. In 1973 Paul Berg – father of genetic engineering invents a method of joining DNA from two different organisms.

Actually Genetic engineering is the process by which scientists modify the genome of an organism. Creation of genetically modified organisms requires recombinant DNA. Recombinant DNA is a combination of DNA from different organisms or different locations in a given genome that would not normally be found in nature.

In most cases, use of recombinant DNA means that you have added an extra gene to an organism to alter a trait or add a new trait. Some uses of genetic engineering include improving the nutritional quality of food, creating pest-resistant crops, and creating infection-resistant livestock.

In other word Genetic engineering, the artificial manipulation, modification, and recombination of DNA or other nucleic acid molecules in order to modify an organism or population of organisms.

Organisms GMO

We're struggling with the fact that some diabetic patients are experiencing an allergic reaction in response to insulin isolated from cows and pigs. The obvious solution is to give patients human insulin instead of animal insulin. Injecting patients with purified human protein should minimize the allergic reaction patients are experiencing, but purifying insulin from human tissue is impractical. That's why cattle or pig insulin is being used at this time.

Let's see how genetic engineering opened the door to mass producing insulin rather than purifying it from animal tissue, human or otherwise.

Historical Developments

The term genetic engineering initially referred to various techniques used for the modification or manipulation of organisms through the processes

Basic Concepts of Genetics

of heredity and reproduction. As such, the term embraced both artificial selection and all the interventions of biomedical techniques, among them artificial insemination, in vitro fertilization (e.g., "test-tube" babies), cloning, and gene manipulation. In the latter part of the 20th century, however, the term came to refer more specifically to methods of recombinant DNA technology (or gene cloning), in which DNA molecules from two or more sources are combined either within cells or in vitro and are then inserted into host organisms in which they are able to propagate.

he possibility for recombinant DNA technology emerged with the discovery of restriction enzymes in 1968 by Swiss microbiologist Werner Arber. The following year American microbiologist Hamilton O. Smith purified so-called type II restriction enzymes, which were found to be essential to genetic engineering for their ability to cleave a specific site within the DNA (as opposed to type I restriction enzymes, which cleave DNA at random sites). Drawing on Smith's work, American molecular biologist Daniel Nathans helped advance the technique of DNA recombination in 1970–71 and demonstrated that type II enzymes could be useful in genetic studies. Genetic engineering based on recombination was pioneered in 1973 by American biochemists Stanley N. Cohen and Herbert W. Boyer, who were among the first to cut DNA into fragments, rejoin different fragments, and insert the new genes into E. coli bacteria, which then reproduced.

A Genetic Revolution

A startling discovery of enormous implications has just been reported in the premier research journal, Science. Despite the accepted dogma that all of a person's cells have the same genetic coding, it turns out that this is not true, especially in neurons. The DNA in each nerve cell (we don't know about sex cells) has hundreds of mutations of the A-T, C-G nucleotides that constitute the genetic code for the neuron. Thus, no two neurons are alike. The study was conducted by 18 research teams at 15 U.S. institutions, formed as a consortium by the National Institute of Mental Health to examine neural genetic coding, using repositories of postmortem brain tissue taken from both healthy people and those with various mental diseases.

The scientists have no explanation at present for what is causing so many mutations and why each neuron has a different genetic profile. The most obvious possibility might seem to be that the mutations occurred as transcription errors during cell division. But there is a major problem with this explanation. We don't know when these mutations occurred. Except for granule cells in the hippocampus and cerebellum, neurons generally do not divide after the first few days after birth. So, if cell division is the cause of mutations, it must be due to what happens during the early post-natal period.

If the mutations occurred sporadically throughout a lifetime, a likely cause for mutation might be DNA damage caused by the free radicals that are generated in ordinary metabolism. Environmental toxins are another possible cause. The point is that the DNA changes are likely to affect how a neuron functions, and that change can last a lifetime.

We know that mutations can cause brain cancer and even certain other brain diseases. The research consortium was commissioned to see if the genetic variants predisposed to neuropsychiatric disease. Obviously, the vast majority of people have these diverse genetic codes in their neurons that do not cause disease. What do they cause? Can the mutations affect which neurons participate in which circuits? Can mutations affect how well you reason, or memorize, or your emotional responsivity? Nobody knows.

A whole new field of research has now been opened. Scientists need to examine different neuronal cell types to see if they are equally affected by mutation. Obvious comparisons needed are between granule cells and all the other neuron types that do not divide.

There is a related aspect that is not considered in this context. That is the likelihood that each neuron differs not only in its genetic code, but also in which genes are expressed. The new field of "epigenetics" has revealed that environmental influences, ranging from drugs, toxins, metabolites, and perhaps even lifestyles can affect the expression of genes, even when there is no mutation. In the case of brain, there is the distinct possibility that one's mental life can affect gene expression. This needs to be studied.

So far, what I have said about gene change and expression refers to single individuals. But what if some of these gene mutations or epigenetic effects that occur in neurons also occur in sex cells? That would mean that traits acquired during one's lifetime could be passed on to future generations. I would hope that the research consortium that has made this monumental discovery about brain cells will extend its charter to also examine sperm and ova.

Recent research on the genetics of the classic animal model of brain function, C. elegans, reveals that epigenetic inheritance of neuronal traits does occur. Gene expression was modified by exposing the animals to high temperatures, and the genetic change was conveyed via both ova and sperm to offspring that had no exposure to high temperature. The epigenetic change was still present some 5-14 generations later.

To the extent that the findings of both of these studies can be extrapolated to humans, we must now consider the possibility that personal lifestyle, environmental, and cultural influences on people may be propagated to successive generations of their children. Bad environments and lifestyle choices may extend well into the future, magnifying the deleterious consequences through multiple generations. We now have to consider that medical and behavioral problems, poverty, and degenerate cultures can arise not only when people make poor choices but that the effects can be genetically propagated to subsequent generations.

These issues may seem to present a challenge to the notion that humans have free will. We are programmed by things that happen to us. But do we not have a choice in deciding much of what we expose ourselves to?

Career in Genetics

"Genetics is the field of medical courses where, we study about heredity, genetic variation & genetic in living beings. It is also known as "Science of Heredity".

Trends in Genetics was launched in 1985 and quickly became a "must read" journal for geneticists, known for its concise, accessible articles on a range of topics from developmental biology to evolution. This tradition continues today, and TiG remains a favorite in the community for its distinctive content. As the field has changed, though, so too has the scope of the journal, which now encompasses new areas, such as genomics, epigenetics, and computational genetics, while continuing to cover traditional subjects like transcriptional regulation, population genetics, and chromosome biology. The core aim of the journal, however, is still to provide researchers and students with high-quality, novel reviews, commentaries, and discussions and, above all, to foster an appreciation for the advances being made on all fronts of genetic research.

Each issue of TiG contains concise, lively and up-to-date Reviews and Opinions as well as a variety of shorter articles, such as Science & Society and Spotlight pieces. Reviews are invited from leading researchers in a specific field and objectively chronicle recent and important developments. Opinon articles provide a forum for debate and hypothesis, and shorter articles discuss aspects of genetics at the intersection of science and policy as well as emerging ideas in the field. All articles are peer-reviewed.

TiG welcomes correspondence. The decision to publish rests with the Editor, and the author(s) of any article discussed in a Letter will normally be invited to reply. Letters may address topics raised in recent issues of the journal, or other matters of general interest to geneticists.

Careers in Human Genetics

At the center of scientific advancement in the 21st century, geneticists are immersed in exciting science, technology, and medical breakthroughs every day. The opportunities are numerous to contribute to the advancement of science, the care of patients, and the teaching of the next generation of genetics professionals. One must be ready to make a commitment in time, energy, and focus to be a successful geneticist, but the rewards are enormous. What more exciting and energizing field could anyone choose?

If you are interested in becoming a geneticist, regardless of your eventual career path, you should start by taking plenty of math and science courses in high school, especially biology and chemistry. In college, biology, chemistry, and biochemistry are popular majors for those interested in genetics careers. Larger institutions may offer more specialized majors such as genetics or molecular biology. Again, you will need to take plenty of math and science classes, and do well in these classes. With your undergraduate science degree, many paths in genetics are open to you!

What is the job demand for geneticists?

As the details of the human genome unfold, the variety of opportunities for people with degrees and training in human genetics is continuing to expand. There are opportunities in basic and clinical research, in medical professions, and in interdisciplinary fields, such as patent law. The genetics workforce is not sufficient even now, and demand continues to increase. For example, as genetic testing becomes more commonplace, and a part of many routine medical evaluations, more laboratory geneticists will be needed to perform the tests, and clinicians and counselors will be needed to interpret and explain the results to individuals and families. At the intersection of genetics and computer science, bioinformaticists are in high demand to make sense of complex data. As genetics is recognized to be a basic part of all biological sciences, more and more teachers with expertise in genetics will also be needed. These are just a few examples of the growing demand for professions trained in genetics.

What is a typical week like?

A geneticist's work week structure and duties vary greatly depending on their career field. Geneticists working in fields such as media, law, public policy, or education follow the typical work week schedule of those fields. Research and health professionals have some control over their work schedules, but often work more than 40 hours every week. For researchers, experiments may take many hours to complete, or require lab work several days in a row. In addition to the laboratory and/or field work required to perform the studies, researchers must read the scientific literature, analyze their own data, and prepare manuscripts of their work for scientific journals. Also, most medical research is quite expensive, and researchers are responsible for competing nationally for funding to support their work by writing successful grant applications. In the case of practicing clinical and laboratory geneticists and genetic counselors, patients come first. That commitment may translate into long days to complete evaluations or follow up, or may include weekend obligations.

What are the salary and benefits?

he salaries and fringe benefits vary, depending on educational background (i.e., highest degree held) and the position taken, but in general, people working in genetics are well-compensated professionals. Geneticists working in university medical centers or research institutions would have salaries and benefits typical of faculty members of similar rank. For those geneticists choosing a career in the private sector (e.g., a biotechnology company), the salary and benefits might vary according to the resources of the company.

What skills, abilities, or personality characteristics should a geneticist possess?

Good geneticists have a basic curiosity and passion about the genetic basis of health and disease. To be successful at the job, one must have perseverance, patience, and good communication skills. Geneticists love to learn and are self-motivated. Students interested in genetics careers should take as much science and math as possible in high school and college.

Limitation of Genetic

Rapid advances in genetic medicine are constantly improving doctors' ability to detect mutations and diagnose disorders. However, genetic testing does have limitations:
- Genetic tests offer only a diagnosis, not a cure.
- Genetic testing cannot always predict the severity of a disease.
- Lab errors are rare but can occur. It is important to choose a reputable testing facility.
- There are still many genetic disorders for which tests have not yet been developed.
- Genetic test results are not straightforward, and interpreting them can be challenging even for a trained medical professional.
- Predictive testing for mutations associated with multifactorial diseases such as cancer won't tell you whether or not you will get the disease. It will only tell you about your genetic risk factors.
- No test is 100% definitive. Accuracy is generally high but varies depending on the disorder. To find out the detection rate for a specific condition, see our dictionary of genetic diseases.

What are the risks and limitations of genetic testing?

The physical risks associated with most genetic tests are very small, particularly for those tests that require only a blood sample or buccal smear (a procedure that samples cells from the inside surface of the cheek). The procedures used for prenatal testing carry a small but real risk of losing the pregnancy (miscarriage) because they require a sample of amniotic fluid or tissue from around the fetus.

Many of the risks associated with genetic testing involve the emotional, social, or financial consequences of the test results. People may feel angry, depressed, anxious, or guilty about their results. In some cases, genetic testing creates tension within a family because the results can reveal information about other family members in addition to the person who is tested. The possibility of genetic discrimination in employment or insurance is also a concern. Genetic testing can provide only limited information about an inherited condition. The test often can't determine if a person will show symptoms of a disorder, how severe the symptoms will be, or whether the disorder will progress over time. Another major limitation is the lack of treatment strategies for many genetic disorders once they are diagnosed.

A genetics professional can explain in detail the benefits, risks, and limitations of a particular test. It is important that any person who is considering genetic testing understand and weigh these factors before making a decision.

Contemporary Genetics – Conclusion

Genetics play a very large role in the appearance and behavior of all organisms. Genetics are the reason for the inheritance of certain traits such as the white forelock or the bent pinky. Most genetic traits are inherited from the person's biological parents, and some people may have a recessive allele present rather than a dominant one. Just because a trait is "dominant", doesnt mean that it will be the trait thats predominantly common in society. Genetics give people their identity, their own personal imprint that makes them unique. Everyone is different.The reason why no one looks the same comes from our variety of genetic composure. Through chance, dominance and recessiveness of genes, people have different genotypes and phenotypes. Even though you can be a family member to someone or even be a twin, your gene makeup can be quite different. Yet, though a certain dominant trait may be commonly expressed among a family, a member can still express a recessive trait due to the fact that we learned - the dominant trait might not always be the most prevalent. An example of this would be the white forelock gene mentioned earlier. It is a dominant gene that is not prevalent

Basic Concepts of Genetics 33

in the human population. There were also instances where the most commonly expressed allele was recessive instead of dominant as most people assume.

Our hypothesis was that the dominant traits are more prevalent than do the recessive traits. However, we concluded that the dominant trait isn't necessarily more prevalent in a population. The frequency of an allele whether it is dominant or recessive depends on how many people in a population carry the allele. So the more prevalent the trait, the higher the frequency. Some dominant traits, white forelock for instance, are very rare in society and simply because it is a dominant trait does not mean it will be seen more frequently in a society. From our data we concluded that 0% of the people we interviewed had the white forelock, disproving our hypothesis that the dominant trait is more prevalent in a population. Most people carry the recessive allele, even though a lot of us assumed that the dominant allele is more prevalent. It shows that the recessive allele can be more prevalent than do the dominant one, thus proving our hypothesis incorrect. The dominant allele actually has a higher chance of being the less commonly expressed allele than that of the recessive allele because of the fact that the recessive allele is normally always carried, it is only when the dominant allele exists in one of the parents that it is shown.

Genetics and the mechanisms of evolution are becoming increasingly important in medicine, as we unintentionally breed strains of highly resistant bacteria. Antibiotic stewardship, or the use of the appropriate antibiotics only as necessary, is very important as the medical community seeks to preserve the effectiveness of antibiotics. In order to understand and apply the concepts of antibiotic stewardship, one must be aware of how creating environmental pressures leads to directional selection in microorganisms and can increase the frequency of the resistant phenotype. In this chapter, we covered genetics and mutations, as well as evolution. We also gave you some tools to analyze the mathematical (statistical) side of genetics through the use of Punnett squares, recombinant frequencies, and the Hardy–Weinberg equations.

New genetic technologies, whether those associated the classical genetics of the first half of the twentieth century or the molecular genetics and genomics of the second half, have al- ways raised a wide variety of ethical issues within the larger society. Whether genetic knowledge is being used politically to place blame for social problems on "defective biology" or genetic engineering technologies are being used to produce "designer babies," geneticists have continually found themselves in the midst of highly controversial issues, ones that are often far more difficult and complex than those associated with other biomedical technologies. This may be in part a result of the long-standing, though mistaken, view that "genetics is destiny" and that knowing the genotype (genetic makeup) of an organism can lead to accurate predictions about its ultimate phenotype (that is, what actual traits will appear and in what form). But it is also in part due to Western society's optimistic faith that science and technology can provide answers to larger economic and social issues. This is an unrealistic view of what role scientific and technological information can play in human life. There is no question that knowing the science involved in any given area of biomedicine (especially human genetics) is critical for making social and political decisions. But it is never enough. Even if scientists could predict with complete accuracy the exact clinical effects that would characterize a fetus with Down's syndrome or Huntington's disease, the decision about how to respond to that knowledge would involve social, political, economic, and philosophical considerations that lie outside of the science itself. As much as anything else, consideration of the ethical and moral aspects of genetic technology should be a reminder that science itself is not, nor has it ever been, a magic bullet for the solution of social problems. Nowhere has that been demonstrated more clearly than in the history of genetics in the twentieth century.

2

Molecular Genetics

INTRODUCTION

Molecular genetics is the field of biology that studies the structure and function of genes at a molecular level and thus employs methods of both molecular biology and genetics. The study of chromosomes and gene expression of an organism can give insight into heredity, genetic variation, and mutations. This is useful in the study of developmental biology and in understanding and treating genetic diseases.

The Molecular Evolutionary Genetics Analysis (MEGA) software is developed for comparative analyses of DNA and protein sequences that are aimed at inferring the molecular evolutionary patterns of genes, genomes, and species over time (Kumar et al. 1994; Tamura et al. 2011). MEGA is currently distributed in two editions: a graphical user interface (GUI) edition with visual tools for exploration of data and analysis results (Tamura et al. 2011) and a command line edition (MEGA-CC), which is optimized for iterative and integrated pipeline analyses (Kumar et al. 2012).

The term molecular genetics is now redundant because contemporary genetics is thoroughly molecular. Genetics is not made up of two sciences, one molecular and one non-molecular. Nevertheless, practicing biologists still use the term. When they do, they are typically referring to a set of laboratory techniques aimed at identifying and/or manipulating DNA segments involved in the synthesis of important biological molecules. Scientists often talk and write about the application of these techniques across a broad swath of biomedical sciences. For them, molecular genetics is an investigative approach that involves the application of laboratory methods and research strategies. This approach presupposes basic knowledge about the expression and regulation of genes at the molecular level.

Philosophical interest in molecular genetics, however, has centered, not on investigative approaches or laboratory methods, but on theory. Early philosophical research concerned the basic theory about the make-up, expression, and regulation of genes. Most attention centered on the issue of theoretical reductionism. The motivating question concerned whether classical genetics, the science of T. H. Morgan and his collaborators, was being reduced to molecular genetics. With the rise of developmental genetics and developmental biology, philosophical attention has subsequently shifted towards critiquing a fundamental theory associated with contemporary genetics. The fundamental theory concerns not just the make-up, expression, and regulation of genes, but also the overall role of genes within the organism. According to the fundamental theory, genes and DNA direct all life processes by providing the information that specifies the development and functioning of organisms.

This article begins by providing a quick review of the basic theory associated with molecular genetics. Since this theory incorporates ideas from the Morgan school of classical genetics, it is useful to sketch its development from Morgan's genetics. After reviewing the basic theory, I examine four questions driving philosophical investigations of molecular genetics. The first question asks whether classical genetics has been or will be reduced to molecular genetics. The second question concerns the gene concept and whether it has outlived its usefulness. The third question regards the tenability of the fundamental theory. The fourth question, which hasn't yet attracted much philosophical attention, asks why so much biological research is centered on genes and DNA.

PROPERTIES of DNA and of The HUMAN GENOME

Genes are made up of deoxyribonucleic acid (DNA). DNA is a long polymer made up of four components, called deoxyribonucleotides. Each deoxyribonucleotide has one of the following constituents, called bases: adenine (A), guanine (G), thymine (T), and cytosine (C). A gene contains a few thousand to a few hundred thousand bases that are strung together in a particular sequence. The base sequence of a gene determines the structure of the gene's product, a protein. Other parts of the gene's base sequence determine the way that the gene is expressed during development of the individual and in response to various stimuli.

DNA's role in storing and transferring hereditary information depends on an inmate property of its four constituent bases. Each of these four bases has structural features which lead it to associate with one, and only one, of the other bases. Two bases that can associate with each other are said to be complementary. Guanine and cytosine are complementary to each other as are adenine and thymine.

Molecular Genetics

Complementary base pairing plays an essential role in maintaining the stability of DNA and also in the transfer of its innate information. Stability is brought about because the DNA in our genes consists of two complementary strands which, by interacting with each other, shield the DNA from perturbations. Information is transferred by separating these two strands, which can then act as templates for the synthesis of new nucleic acid molecules.

DNA molecules may be used as templates in two critical ways. First the DNA is used as a template for replicating additional copies of DNA, which is essential for cell division. In this case, free deoxyribonucleotides bond with exposed complementary bases in each of the two template strands and are then linked together by an enzyme, DNA polymerase. The product is two new complementary chains which, together, reproduce the template. In DNA replication, complete DNA strands, made up of tens of millions of bases, are copied—so they can be transmitted to new cells. The other way that DNA is used as a template is more selective and has a different purpose. In this case, small bits of a strand are used as a template for the construction of molecules called messenger RNAs (mRNAs), each of which carries the message for the synthesis of a particular protein.

Messenger RNAs differ from DNA in a number of ways. First they are much shorter—generally on the order of a thousand to several thousand base pairs long. Second, they are made up of only a single strand (in contrast with double-stranded DNA) that contains all the requisite information to direct the synthesis of a particular protein. They also have a different sugar in their nucleic acid backbone—ribose rather than deoxyribose; and the thymine found in DNA is replaced with a similar base, uracil (U) in mRNA. Like thymine, uracil is complementary to adenine.

The entire human genome consists of about 100,000 genes distributed within a total DNA sequence of about 3 billion nucleotides. The DNA of the human genome is divided into 24 huge molecules, each the essential constituent of a particular chromosome (22 autosomes and two different sex chromosomes, X and Y). When we are conceived we receive 23 chromosomes from each parent, 22 autosomes and a sex chromosome.

As already indicated, a major function of a gene is to encode the structure of a specific protein. Translation of the information encoded in DNA (which is expressed in an alphabet of nucleotides) into a protein (which is expressed in an alphabet of amino acids) depends on a genetic code. In this code, sequences of three nucleotides, called a codon, represent one of the 20 amino acids that comprise the building blocks of all proteins. Because there are 64 possible codons that can be constructed from an alphabet consisting of four different bases, and only 20 different amino acids to be coded for, many amino acids are encoded by more than one particular codon. Three of the codons, called stop codons, are used to signal termination of translation.

Although all cells express certain genes that are required for their shared housekeeping functions, the distinctive differences between specialized cells (such as particular classes of neurons) are due to selective gene expression. For example, in the nervous system certain neurons use acetylcholine and others use dopamine as neurotransmitters. This results from selective expression of genes that encode specific proteins (in this case enzymes) that catalyze the biosynthesis of either acetylcholine or dopamine. Expression of these genes is ultimately under the control of specific regulatory proteins, called transcription factors, that bind to regions of the genes. These regulatory proteins control the transcription of mRNA from the genes they control. The regulatory proteins that control specific genes are, themselves, selectively expressed during the maturation of the particular classes of neurons.

Expression of enzymes that control biosynthesis of neurotransmitters is controlled not only by factors that operate during embryonic development, but also by factors that influence the adult organism. For example, the synthesis of certain of these enzymes depends, in part, on neuronal activation. When there is more neuronal activation, more of a critical enzyme is made and, as a result, the neuron makes more of the neurotransmitter. Complex regulation of genes involved in neurotransmitter biosynthesis (and of other genes for receptors and transporters that determine neurotransmitter function) may play important roles in the control of behavior. Regulation of genes also determines the response of the nervous system to drugs, the central concern of this volume.

The study of these genetic and cellular regulatory processes is one of the most active areas of contemporary biology. At present a great deal is being learned about the specific base sequences, called regulatory sequences, that surround the portions of the gene dedicated to encoding the sequence of a protein. These sequences are activated or inactivated by the specific transcription factors that bind to and control them. The complex interaction of regulatory sequences and transcription factors underlies adaptation of the brain to drugs. In the case of antidepressants and neuroleptics, these adaptive changes are essential for the therapeutic effect, which only develops after weeks of drug treatment. Because adaptive changes are essential, understanding them should lead to the design of new psychopharmacological agents

MANIPULATING DNA

One of the ultimate goals of molecular genetics is to determine the exact base sequence of all 3 billion bases that comprise the human genome. This task is very challenging because DNA molecules are gigantic, which makes them extremely difficult to deal with.

A major step toward achieving this goal came from development of methods to isolate, and then examine, the bite-size pieces of DNA that had been transcribed into mRNAs. The information encoded into mRNAs can be isolated and amplified by a technique called cDNA cloning Figure 1 . In this technique, a mixture, containing all the mRNAs from an organ, such as the brain, are first purified. The mixture of mRNAs are then treated with an enzyme, called reverse transcriptase, which transcribes the mRNAs into single complementary strands of DNA called complementary DNA (cDNA). Single-stranded cDNAs are then used as templates to make a second strand that is complementary to the first; and the double-stranded cDNAs are then inserted into bacterial plasmids to make products called recombinant DNA plasmids. The plasmids are then inserted into specially engineered bacteria in which they are replicated, along with the bacterial DNA, during the process of bacterial cell division. In this way many copies of the cDNAs are made. The bacterial population is comprised of many individual bacteria, each of which contains a particular plasmid with a particular cDNA derived from an mRNA from the original tissue sample. This mixed population is called a cDNA library.

To physically separate the individual members of the library, the bacteria are grown on a solid nutrient agar at low density. Each bacterium in the library is plated onto the agar at a large distance from the others. As each bacterium divides on the agar, it gives rise to a colony of descendants called a clone, which is physically separate from other clones derived from other bacteria that contain other cDNAs. Each member of a clone carries copies of the cDNA-containing plasmid that had been inserted into the clone's founder. Each clone may then be separately removed from the agar (without contamination with bacteria from other clones), and bacteria that all contain the particular cDNA may be grown up in large quantities. Then the cDNA within the plasmids can be excised, and its nucleotide sequence can be determined by chemical techniques. The cloned cDNA may also be used for many other purposes, a few of which are discussed later.

Of course the brain expresses many mRNAs that are also expressed by other tissues and that may have no special interest for neurobiological or pharmacological research. In most studies of brain cDNA the goal is to find the one that encodes the sequence of a particular protein of interest, such as a receptor protein for a particular neurotransmitter. There are many ways to go about searching for a specific cDNA (and its specific clone). One involves insertion of cDNA-containing plasmids (derived from bacterial clones) into cultured mammalian cells (such as fibroblasts) that can express the neurotransmitter receptor on their cell surface (in contrast with bacteria which do not process the cDNAs in the same way). The cDNA of interest is sought by reacting the mammalian cell population with a ligand (such as a neurotransmitter or a drug that binds the receptor that is being sought) and isolating the cells that bind the ligand.

Once a particular cDNA is isolated, it can be used to make limitless quantities of the protein whose sequence it encodes. For some proteins, this can be done in bacteria. In this case the plasmid can be induced to make mRNA that is translated by the bacterial protein-synthesizing machinery. However, in many cases the translation is done in mammalian cells, so that the protein product not only has the amino acids sequence encoded by the cDNA, but also undergoes appropriate post-translational modifications, such as glycosylation, which do not occur with expression in bacterial cells. In the case of neurotransmitter receptors the desired product may not be a pure soluble receptor protein, but may instead be a mammalian population of cells that express the receptor as a protein integrated into the plasma membrane on the cell surface. Cells with a particular receptor on their surface can then be used to screen for drugs that bind this receptor.

Cloned cDNA can also be used as a reagent in a variety of biological and medical studies. These are generally based on the innate property of nucleic acids to undergo complementary base paring, so that a single-stranded cDNA will bind to complementary nucleic acid sequences in mixtures of human nucleic acid, or even in tissue sections, by a process called hybridization. If the cDNA probe has been prepared in a radioactive form, the amount of radioactive cDNA that hybridizes to an aliquot of a tissue extract provides a measure of the amount of the mRNA that is complementary to it in the tissue extract. The radioactive cDNA probe can also be applied to brain tissue sections to localize the mRNA in specific neuronal populations in the brain, by a process called in situ hybridization. In this way the distribution of a particular protein, such as a receptor, can be inferred by determining the distribution of the mRNA that codes for this protein. The distribution of a particular receptor may have important implications for the design of drugs that are targeted to a specific brain region.

MANIPULATING GENES

Once it became possible to isolate cDNAs that encode proteins of particular interest, a variety of techniques were developed to use the cDNAs to learn about the function of the proteins. The basic idea is to insert the cDNAs (modified by the addition of regulatory sequences that make possible their controlled expression; and, at times, also modified in other ways) into cells, then measure specific effects. Some of these studies are done in cells in tissue culture, whereas others are done by changing the genetic composition of intact organisms. This type of gene manipulation underlies very powerful approaches to the study of the function of given proteins, and it also provides cell types and animals with many applications in psychopharmacological research.

The simplest manipulation of this type is to introduce a new gene (or many additional copies of a particular gene) into a cultured cell line, a process called transfection. This is accomplished by engineering the cDNA into various vectors, such as appropriate plasmids, that will carry it into the cell and allow it to be expressed. In one form of transfection, called stable transfection, the cDNA (along with regulatory sequences) or other type of foreign DNA is stably integrated into the DNA of a chromosome. When a cell of this type replicates its DNA for cell division, the integrated cDNA is also replicated and transmitted to the daughter cells. To obtain stably transfected cells it is necessary to select them from a population that consists largely of cells that have not integrated the cDNA of interest.

The desired cells may be selected from a mixed population that also consists of many cells that do not contain the desired cDNA by a trick of genetic engineering. The trick is to transfect the cDNA of interest along with other DNA that makes possible the survival of transfected cells under experimentally induced toxic conditions. For example, the transfected DNA may include a sequence for a protein that renders the recipient cell resistant to a toxic compound so that only cells containing the transfected DNA (including the cDNA of interest) will survive if this toxic compound is added to the culture medium. In this way, clones of cells that contain the cDNA of interest can be isolated and used for various purposes.

To make animals that express a particular segment of foreign DNA, it is possible to inject this particular DNA into one-celled embryos. The DNA is then integrated into the genome of the recipient embryo and into all its cells, including its germ cells, so that it will be transmitted to future generations. Most of these experiments are done with mice, and a single mouse with a segment of foreign DNA incorporated into its genome can give rise to a line of mice, each of which has the foreign DNA, which is called a transgene.

Such mice are called transgenic mice. Regulatory sequences surrounding the coding sequence of the transgene may bring it under specific control. For example, certain regulatory sequences will direct expression of the transgene only in a particular cell type, such as muscle cells.

A particularly interesting variety of transgenic mice has a foreign gene inserted not in addition to, but in place of, a normal gene. This is accomplished by a technique in which a normal gene is removed from a chromosome in the same process in which the foreign gene is inserted. One common application of this approach is to replace a normal gene in the germ line with one that is inactive or defective, thereby giving rise to progeny that lack the normal gene and its function. By mating brothers and sisters each with one defective gene copy, progeny can be raised that have two defective genes (i.e., no normal copies of the gene). Such "knockout experiments" are one way of examining the normal biological role of the gene in question.

In some cases the results are not very informative because absence of the functional gene during early embryonic development results in the death of the embryo. In other cases, loss of a particular gene has no obvious effect, presumably because other genes take over for the one that was inactivated. In many cases, however, mice that lack a particular gene have proved useful for determining a particular gene's function. In the context of the present volume, mice lacking a particular receptor for a neurotransmitter may provide clues to this receptor's function in the cells that express it, and in the animal as a whole. Animals lacking a particular receptor may also prove useful for certain approaches to drug development.

What is a gene?

A common claim in the philosophical literature about molecular genetics is that genes cannot be conceived at the molecular level. Of course, philosophers do not deny that biologists use the term gene, but many philosophers believe gene is a dummy term, a placeholder for many different concepts. Different responses to gene skepticism illustrate a variety of philosophical aims and approaches. One kind of response is to analyze explanations closely tied to experimental practice (rather than sweeping generalizations of a fundamental theory) in order to determine whether there are uniform patterns of reasoning about genes that could (a) be codified into clear concepts, and/or (b) used to establish the reference of the term. Another kind of response is to propose new gene concepts that will better serve the expressed aims of practicing biologists. A third kind of response is to implement survey analysis, rather than conduct traditional methods of philosophical analysis. A fourth kind of response is to embrace the (allegedly) necessary vagueness of the gene concept(s) and to examine why use of the term gene is so useful.

Gene skepticism

Gene skeptics claim that there is no coherence to the way gene is used at the molecular level and that this term does not designate a natural kind; rather, gene is allegedly used to pick out many different kinds of units in DNA. DNA consists of "coding" regions that are transcribed into RNA, different kinds of regulatory regions, and in higher organisms, a number of regions whose functions are less clear and perhaps in cases non-existent. Skepticism about genes is based in part on the idea that the term is sometimes applied to only parts of a coding region, sometimes to an entire coding region, sometimes to parts of a coding region and to regions that regulate that coding region, and sometimes to an entire coding region and regulatory regions affecting or potentially affecting the transcription of the coding region. Skeptics (e.g., Burian 1986, Portin 1993, and Kitcher 1992) have concluded, as Kitcher succinctly puts it: "a gene is whatever a competent biologist chooses to call a gene."

Biological textbooks contain definitions of gene and it is instructive to consider one in order to show that the conceptual situation is indeed unsettling. The most prevalent contemporary definition is that a gene is the fundamental unit that codes for a polypeptide. One problem with this definition is that it excludes many segments that are typically referred to as genes. Some DNA segments code for functional RNA molecules that are never translated into polypeptides. Such RNA molecules include transfer RNA, ribosomal RNA, and RNA molecules that play regulatory and catalytic roles. Hence, this definition is too narrow.

Another problem with this common definition is that it is based on an overly simplistic account of DNA expression. According to this simple account, a gene is a sequence of nucleotides in DNA that is transcribed into a sequence of nucleotides making up a messenger RNA molecule that is in turn translated into sequence of amino acids that forms a polypeptide. (Biologists talk as if genes "produce the polypeptide molecules" or "provide the information for the polypeptide".) The real situation of DNA expression, however, is often far more complex. For example, in plants and animals, many mRNA molecules are processed before they are translated into polypeptides. In these cases, portions of the RNA molecule, called introns, are snipped out and the remaining segments, called exons, are spliced together before the RNA molecule leaves the cellular nucleus. Sometimes biologists call the entire DNA region, that is the region that corresponds to both introns and exons, the gene. Other times, they call only the portions of the DNA segment corresponding to the exons the gene. (This means that some DNA segments that geneticists call genes are not continuous segments of DNA; they are collections of discontinuous exons. Geneticists call these split genes.) Further complications arise because the splicing of exons in some cases is executed differentially in different tissue types and at different developmental stages. (This means that there are overlapping genes.) The problem with the common definition that genes are DNA segments that "code for polypeptides" is that the notion of "coding for a polypeptide" is ambiguous when it comes to actual complications of DNA expression. Gene skeptics argue that it is hopelessly ambiguous (Burian 1986, Fogle 1990 and 2000, Kitcher 1992, and Portin 1993).

Clearly, this definition, which is the most common and prominent textbook definition, is too narrow to be applied to the range of segments that geneticists commonly call genes and too ambiguous to provide a single, precise partition of DNA into separate genes. Textbooks include many definitions of the gene. In fact, philosophers have often been frustrated by the tendency of biologists to define and use the term gene in a number of contradictory ways in one and the same textbook. After subjecting the alternative definitions to philosophical scrutiny, gene skeptics have concluded that the problem isn't simply a lack of analytical rigor. The problem is that there simply is no such thing as a gene at the molecular level. That is, there is no single, uniform, and unambiguous way to divide a DNA molecule into different genes. Gene skeptics have often argued that biologists should couch their science in terms of DNA segments such exon, intron, promotor region, and so on, and dispense with the term gene altogether.

Basic Theory of Genetics

The basic theory of genetics is divided in two part as described below

1 The basic theory of classical genetics

The basic theory associated with classical genetics provided explanations of the transmission of traits from parents to offspring. Morgan and his collaborators drew upon a conceptual division between the genetic makeup of an organism, termed its genotype, and its observed manifestation called its phenotype (see the entry on the genotype/phenotype distinction). The relation between the two was treated as causal: genotype in conjunction with environment produces phenotype. The theory explained the transmission of phenotypic differences from parents to offspring by following the transmission of gene differences from generation to generation and attributing the presence of alternative traits to the presence of alternative forms of genes.

I will illustrate the classical mode of explanatory reasoning with a simple historical example involving the fruit fly Drosophila melanogaster. It is worth emphasizing that the mode of reasoning illustrated by this historical

Molecular Genetics

example is still an important mode of reasoning in genetics today, including what is sometimes called molecular genetics.

Genes of Drosophila come in pairs, located in corresponding positions on the four pairs of chromosomes contained within each cell of the fly. The eye-color mutant known as purple is associated with a gene located on chromosome II. Two copies of this gene, existing either in mutated or normal wild-type form, are located at the same locus (corresponding position) in the two second-chromosomes. Alternative forms of a gene occurring at a locus are called alleles. The transmission of genes from parent to offspring is carried out in a special process of cellular division called meiosis, which produces gamete cells containing one chromosome from each paired set. The half set of chromosomes from an egg and the half set from a sperm combine during fertilization, which gives each offspring a copy of one gene from each gene pair of its female parent and a copy of one gene from each gene pair of its male parent.

Explanations of the transmission of traits relate the presence of alternative genes (genotype) to the presence of alternative observable traits (phenotype). Sometimes this is done in terms of dominant/recessive relations. Purple eye-color, for example, is recessive to the wild-type character (red eye-color). This means that flies with two copies of the purple allele (the mutant form of the gene, which is designated pr) have purple eyes, but heterozygotes, flies with one copy of the purple allele and one copy of the wild-type allele (designated +), have normal wild-type eyes (as do flies with two copies of the wild-type allele).

To see how the classical theory explains trait transmission, consider a cross of red-eyed females with purple-eyed males that was carried out by Morgan's collaborators. The offspring all had red eyes. So the trait of red eyes was passed from the females to all their offspring even though the offspring's male parents had purple eyes. The classical explanation of this inheritance pattern proceeds, as do all classical explanations of inheritance patterns, in two stages.

The first stage accounts for the transmission of genes and goes as follows (Figure 1): each offspring received one copy of chromosome II from each parent. The maternally derived chromosomes must have contained the wild-type allele (since both second-chromosomes of every female parent used in the experiment contained the wild-type allele -- this was known on the basis of previous experiments). The paternally derived chromosomes must have contained the purple allele (since both second-chromosomes of every male parent contained the purple allele -- this was inferred from the knowledge that purple is recessive to red eye-color). Hence, all offspring were heterozygous (pr/ +). Having explained the genetic makeup of the progeny by tracing the transmission of genes from parents to offspring, we can proceed to the second stage of the explanation: drawing an inference about phenotypic appearances. Since all offspring were heterozygous (pr / +), and since purple is recessive to wild-type, all offspring had red eye-color (the wild-type character).

Notice that the reasoning here does not depend on identifying the material make-up, mode of action, or general function of the underlying gene. It depends only on the ideas that copies of the gene are distributed from generation to generation and that the difference in the gene (i.e., the difference between pr and +), whatever this difference is, causes the phenotypic difference. The idea that the gene is the difference maker needs to be qualified: differences in the gene cause phenotypic differences in particular genetic and environmental contexts. This idea is so crucial and so often overlooked that it merits articulation as a principle (Waters 1994):

Difference principle: differences in a classical gene cause uniform phenotypic differences in particular genetic and environmental contexts.

It is also worth noting that the difference principle provides a means to explain the transmission of phenotypic characteristics from one generation to the next without explaining how these characteristics are produced in the process of an organism's development. This effectively enabled classical geneticists to develop a science of heredity without answering questions about development.

The practice of classical genetics included the theoretical analysis of complicated transmission patterns involving the recombination of phenotypic traits. Analyzing these patterns yielded information about the basic biological processes such as chromosomal mechanics as well as information about the linear arrangement of genes in linkage groups. These theoretical explanations did not depend on ideas about what genes are, how genes are replicated, what genes do, or how differences in genes bring about differences in phenotypic traits.

2 Molecular-level answers to questions left behind by classical genetics

Research in molecular biology and genetics has yielded answers to the basic questions left unanswered by classical genetics about the make-up of genes, the mechanism of gene replication, what genes do, and the way that gene differences bring about phenotypic differences. These answers are couched in terms of molecular level phenomena and they provide much of the basictheory associated with molecular genetics.

What is a gene?

This question is dealt with at further length in section 4 of this article, but a quick answer suffices for present purposes: genes are linear sequences

Molecular Genetics

of nucleotides in DNA molecules. Each DNA molecule consists of a double chain of nucleotides. There are four kinds of nucleotides in DNA: guanine, cytosine, thymine, and adenine. The pair of nucleotide chains in a DNA molecule twist around one another in the form of a double helix. The two chains in the helix are bound by hydrogen bonds between nucleotides from adjacent chains. The hydrogen bonding is specific so that a guanine in one chain is always located next to cytosine in the adjacent chain (and vice-versa) and thymine in one chain is always located next to adenine (and vice-versa). Hence, the linear sequence of nucleotides in one chain of nucleotides in a DNA molecule is complimentary to the linear sequence of nucleotides in the other chain of the DNA molecule. A gene is a segment of nucleotides in one of the chains of a DNA molecule. Of course, not every string of nucleotides in DNA is a gene; segments of DNA are identified as genes according to what they do (see below).

How are genes replicated?

The idea that genes are segments in a DNA double helix provides a straightforward answer to this question. Genes are faithfully replicated when the paired chains of a DNA molecule unwind and new chains are formed along side the separating strands by the pairing of complementary nucleotides. When the process is complete, two copies of the original double helix have been formed and hence the genes in the original DNA molecule have been effectively replicated.

What do genes do?

Roughly speaking, genes serve as templates in the synthesis of RNA molecules. The result is that the linear sequence of nucleotides in a newly synthesized RNA molecule corresponds to the linear sequence of nucleotides in the DNA segment used as the template. Different RNA molecules play different functional roles in the cell, and many RNA molecules play the role of template in the synthesis of polypeptide molecules. Newly synthesized polypeptides are linear sequences of amino acids that constitute proteins and proteins play a wide variety of functional roles in the cell and organism (and environment). The ability of a polypeptide to function in specific ways depends on the linear sequence of amino acids of which it is formed. And this linear sequence corresponds to the linear sequence of triplets of nucleotides in RNA (codons), which in turn corresponds to the linear sequence of nucleotides in segments of DNA, and this latter segment is the gene for that polypeptide.

How do differences in genes bring about differences in phenotypic traits?

The modest answer given above to the question 'What do genes do?' provides the basis for explaining how differences in genes can bring about differences in phenotypic traits. A difference in the nucleotide sequence of a gene will result in the difference in the nucleotide sequence of RNA molecules, which in turn can result in a difference in the amino acid sequence of a polypeptide. Differences in the linear sequences of amino acids in polypeptides (and in the linear sequences of nucleotides in functional RNA molecules) can affect the roles they play in the cell and organism, sometimes having an effect that is observable as a phenotypic difference. The mutations (differences in genes) identified by the Morgan group (e.g., the purple mutation) have been routinely identified as differences in nucleotide sequences in DNA.

3 Distinguishing between basic and fundamental theories of molecular genetics

The modest answer to the question 'What do genes do?' is that they "code for" or "determine" the linear sequences in RNA molecules and polypeptides synthesized in the cell. (Even this modest answer needs to be qualified because RNA molecules are often spliced and edited in ways that affect the linear sequence of amino acids in the eventual polypeptide product.) But biologists have offered a far less modest answer as well. The bolder answer is part of a sweeping, fundamental theory. According to this theory, genes are "fundamental" entities that "direct" the development and functioning of organisms by "producing" proteins that in turn regulate all the important cellular processes. It is often claimed that genes provide "the information", "the blueprint", or "the program" for an organism. It is useful to distinguish this sweeping, fundamental theory about the allegedly fundamental role of genes from the modest, basic theory about what genes do with respect to the synthesis of RNA and polypeptides.

Mendel's Genetics

For thousands of years farmers and herders have been selectively breeding their plants and animals to produce more useful hybrids. It was somewhat of a hit or miss process since the actual mechanisms governing inheritance were unknown. Knowledge of these genetic mechanisms finally came as a result of careful laboratory breeding experiments carried out over the last century and a half.

By the 1890's, the invention of better microscopes allowed biologists to discover the basic facts of cell division and sexual reproduction. The focus of genetics research then shifted to understanding what really happens in the transmission of hereditary traits from parents to children. A number of hypotheses were suggested to explain heredity, but Gregor Mendel, a little known Central European monk, was the only one who got it more or less right. His ideas had been published in 1866 but largely went unrecognized until 1900, which was long after his death. His early adult life was spent in relative obscurity doing basic genetics research and teaching high school mathematics, physics, and Greek in Brno (now in the Czech Republic). In his later years, he became the abbot of his monastery and put aside his scientific work.

While Mendel's research was with plants, the basic underlying principles of heredity that he discovered also apply to people and other animals because the mechanisms of heredity are essentially the same for all complex life forms.

Through the selective cross-breeding of common pea plants (Pisum sativum) over many generations, Mendel discovered that certain traits show up in offspring without any blending of parent characteristics. For instance, the pea flowers are either purple or white--intermediate colors do not appear in the offspring of cross-pollinated pea plants. Mendel observed seven traits that are easily recognized and apparently only occur in one of two forms:

1. flower color is purple or white
2. flower position is axil or terminal
3. stem length is long or short
4. seed shape is round or wrinkled
5. seed color is yellow or green
6. pod shape is inflated or constricted
7. pod color is yellow or green

This observation that these traits do not show up in offspring plants with intermediate forms was critically important because the leading theory in biology at the time was that inherited traits blend from generation to generation. Most of the leading scientists in the 19th century accepted this "blending theory." Charles Darwin proposed another equally wrong theory known as "pangenesis". This held that hereditary "particles" in our bodies are affected by the things we do during our lifetime. These modified particles were thought to migrate via blood to the reproductive cells and subsequently could be inherited by the next generation. This was essentially a variation of Lamarck's incorrect idea of the "inheritance of acquired characteristics."

Mendel picked common garden pea plants for the focus of his research because they can be grown easily in large numbers and their reproduction can be manipulated. Pea plants have both male and female reproductive organs. As a result, they can either self-pollinate themselves or cross-pollinate with another plant. In his experiments, Mendel was able to selectively cross-pollinate purebred. plants with particular traits and observe the outcome over many generations. This was the basis for his conclusions about the nature of genetic inheritance.

In cross-pollinating plants that either produce yellow or green pea seeds exclusively, Mendel found that the first offspring generation (f1) always has yellow seeds. However, the following generation (f2) consistently has a 3:1 ratio of yellow to green.

He came to three important conclusions from these experimental results:

1. that the inheritance of each trait is determined by "units" or "factors" that are passed on to descendents unchanged (these units are now called genes)
2. that an individual inherits one such unit from each parent for each trait
3. that a trait may not show up in an individual but can still be passed on to the next generation.

It is important to realize that, in this experiment, the starting parent plants were homozygous for pea seed color. That is to say, they each had two identical forms (or alleles) of the gene for this trait--2 yellows or 2 greens. The plants in the f1 generation were all heterozygous. In other words, they each had inherited two different alleles--one from each parent plant. It becomes clearer when we look at the actual genetic makeup, or genotype, of the pea plants instead of only the phenotype, or observable physical characteristics.

Note that each of the f1 generation plants (shown above) inherited a Y allele from one parent and a G allele from the other. When the f1 plants breed, each has an equal chance of passing on either Y or G alleles to each offspring.

With all of the seven pea plant traits that Mendel examined, one form appeared dominant over the other, which is to say it masked the presence of the other allele. For example, when the genotype for pea seed color is YG (heterozygous), the phenotype is yellow. However, the dominant yellow allele does not alter the recessive green one in any way. Both alleles can be passed on to the next generation unchanged.

Mendel's observations from these experiments can be summarized in two principles:

1. the principle of segregation
2. the principle of independent assortment

Molecular Genetics

According to the principle of segregation, for any particular trait, the pair of alleles of each parent separate and only one allele passes from each parent on to an offspring. Which allele in a parent's pair of alleles is inherited is a matter of chance. We now know that this segregation of alleles occurs during the process of sex cell formation (i.e., meiosis).

Segregation of alleles in the production of sex cells

According to the principle of independent assortment, different pairs of alleles are passed to offspring independently of each other. The result is that new combinations of genes present in neither parent are possible. For example, a pea plant's inheritance of the ability to produce purple flowers instead of white ones does not make it more likely that it will also inherit the ability to produce yellow pea seeds in contrast to green ones. Likewise, the principle of independent assortment explains why the human inheritance of a particular eye color does not increase or decrease the likelihood of having 6 fingers on each hand. Today, we know this is due to the fact that the genes for independently assorted traits are located on different chromosomes These two principles of inheritance, along with the understanding of unit inheritance and dominance, were the beginnings of our modern science of genetics. However, Mendel did not realize that there are exceptions to these rules. Some of these exceptions will be explored in the third section of this tutorial and in the Synthetic Theory of Evolution tutorial.

By focusing on Mendel as the father of genetics, modern biology often forgets that his experimental results also disproved Lamarck's theory of the inheritance of acquired characteristics described in the Early Theories of Evolution tutorial. Mendel rarely gets credit for this because his work remained essentially unknown until long after Lamarck's ideas were widely rejected as being improbable.

Principles of molecular genetic

The role of genetic and genomic information in the practice of clinical medicine is increasing at a rapid pace. Examples include advances in the scope of prenatal screening, including testing of maternal blood for disorders found in the fetus, diagnosis and molecular classification of rare genetic disease using next-generation sequencing, tumor classification by gene expression analysis, and pharmacogenetic applications for medication dosing. Gene identification for complex traits is also progressing rapidly. Findings from these investigations will undoubtedly translate into changes in clinical diagnostics in due time.

Physicians involved in the care of patients will require knowledge of the basic principles of genetics to adequately incorporate these applications into clinical practice. The basic principles of molecular genetics are reviewed here. The material summarized here is essential to understand topics related to the basic science and clinical applications of genetics addressed elsewhere within UpToDate. Readers are encouraged to consult introductory texts in molecular genetics and biology for more detailed reviews of these concepts. Examples of suitable texts are provided in the reference section

The fundamental processes by which heritable information is stored, transmitted from generation to generation, and translated from genetic code to function in living organisms are common to all eukaryotes. These processes were termed the "Central Dogma of Molecular Biology" by Francis, although the use of the word "Dogma" has been debated subsequently.

Heritable genetic information is stored as long stretches of deoxyribonucleic acid (DNA), a stable molecule that is replicated and transmitted from one generation to the next. Segments of DNA can encode functional elements called genes that are transcribed into messenger ribonucleic acid (messenger RNA or mRNA). mRNA then serves as the template for protein synthesis in a process called translation.

mRNA isolation

Expressed DNA that codes for the synthesis of a protein is the final goal for scientists and this expressed DNA is obtained by isolating mRNA (Messenger RNA).

First, laboratories use a normal cellular modification of mRNA that adds up to 200 adenine nucleotides to the end of the molecule (poly(A) tail). Once this has been added, the cell is ruptured and its cell contents are exposed to synthetic beads that are coated with thymine string nucleotides. Because Adenine and Thymine pair together in DNA, the poly(A) tail and synthetic beads are attracted to one another, and once they bind in this process the cell components can be washed away without removing the mRNA. Once the mRNA has been isolated, reverse transcriptase is employed to convert it to single-stranded DNA, from which a stable double-stranded DNA is produced using DNA polymerase. Complementary DNA (cDNA) is much more stable than mRNA and so, once the double-stranded DNA has been produced it represents the expressed DNA sequence scientists look for.[4]

ISOLATION OF mRNA FROM TOTAL RNA

TRIzol homogenization: Total RNA includes all mRNA, transfer RNA, ribosomal RNA, and other noncoding RNAs. To separate these from other cellular components, the cell is first burst open to release its contents. This is done by resuspending cells pelleted by centrifuging (spinning at high

Molecular Genetics

speeds) in TRIzol Reagent (Life Technologies). Other versions of TRIzol (such as Ambion's TRI Reagent) work similarly.

Total RNA Isolation: A series of centrifugations is used to separate the different components (proteins, DNA, RNA) of the cell into layers, or phases, within the suspension. The top, yellow-colored phase is composed of fat and can be discarded. The desired phase is tinted red, contains the total RNA and is retained. After performing a phenol-chloroform extraction and a series of alcohol washes using isopropanol and ethanol, the RNA can be pelleted for mRNA isolation. Add RNase inhibitors to prevent this enzyme from degrading the total RNA.

mRNA Extraction: It is common to use a kit to isolate mRNAs, as homemade lab protocols do not generate large quantities of highly purified mRNAs. Commercial kits include Invitrogen's FastTrack 2.0 or Ambion's Poly(A) Pure mRNA Isolation Kit. These basic steps are common to such kits:

a) Mix the RNase-inhibited lysis buffer provided in the kit with up to 300 microliters of total RNA.
b) Heat for 5 minutes at 65 degrees Celsius and then immediately cool the sample on ice for one minute.
c) Mix this with 0.5M Sodium Chloride and then completely dissolve Oligo dT (oligodeoxythymidylic acid) in this sample.
d) Centrifuge this sample and recover the supernatant, which is washed several times in a series of binding and low salt buffers provided in the kits.
e) Elute mRNA several times until a kit-specified volume (e.g. 500 microliters) is obtained.
f) Precipitate the eluate by sodium acetate and ethanol precipitation. Re-suspend in up to 20 microliters of diethylpyrocarbonate (DEPC)-treated water.
g) Store at -80 degrees Celsius and check for quality and quantity by spectrophotometry.

Tip

Keep all reagents, cells and RNA cold by submerging in ice. This prevents the RNA from being degraded by any other enzymes that become released during the homogenization process.

Warning

Reagents such as TRIzol are toxic and must not be in contact with skin or mucous membranes. Always observe safe lab protocols when handling this reagent.

How mRNA Isolate

mRNAs (messenger RNAs) comprise only a small percentage of all RNA species in a eukaryotic cell, in Neurospora usually ~ 1-6 % (Lucas et al., 1977; Sturani et al., 1979). For some applications like preparation of a cDNA library, target preparation for microarray hybridizations or Northern blot analysis of weakly expressed genes, enriched mRNA preparations are preferable to total RNA. Enrichment of eukaryotic mRNAs derived from nuclear encoded genes is done by virtue of their poly(A) tail which in most cases is 30-200 nt long. mRNA or poly(A)-RNA preparation consists of three steps: (1) hybridization of poly(A)-containing RNAs to oligo-dT molecules connected to a carrier, (2) washing off nucleic acids which do not bind to oligo-dT, (3) elution of poly(A)-RNA from the oligo-dT/carrier combination under low stringency conditions. Procedure: It is most important for isolation of intact, full length mRNAs to keep an RNase-free environment. All reaction tubes, pipet tips, solutions, etc. used to handle (m)RNA should be RNase-free. Glassware can be baked at 180 °C for 6 h, plastic disposables and solutions should be autoclaved twice. Solutions which cannot be autoclaved should be prepared in sterilized vessels with sterilized water and filter sterilized. Wear gloves throughout. Poly(A)-RNA preparation can be done using cellulose-bound oligo-dT (Aviv and Leder, 1972; Chirgwin et al., 1979), but several other carriers for the oligo-dT molecules have been developed, e.g. streptavidin-coupled magnetic beads used in combination with biotinylated oligo-dT or oligo-dT-coupled polystyrene-latex beads. The oligo-dT/carrier combinations are available separately from several manufacturers, but it is more convenient to use a kit which has the advantage of containing most of the necessary reagents pre-packaged in RNase-free quality (e.g. polyATtract from Promega or Oligotex mRNA kit from Qiagen work well for Neurospora, Table 1). Depending on the kit size, this allows isolation of poly(A)-RNA from 0.25 to 5 mg of total RNA. For most kits, total RNA used for mRNA preparation should have a concentration of $= 2$ µg/µl. Total RNA can be prepared with any of the commonly used protocols. One should keep in mind, though, that RNA prepared with a standard phenol/chloroform extraction procedure usually also contains genomic DNA so that the amount of actual RNA in the solution is lower than calculated from OD measurements. Total RNA should be checked by gel electrophoresis and Northern blot for the absence of degradation before using it for mRNA preparation. Visible downward smear from the rRNA bands on a gel or a hybridizing band in a Northern blot indicates degradation of RNA and such preparations should not be used for poly(A)-RNA preparation. mRNA isolation kits usually incorporate the three steps mentioned above. Here, only a brief outline of the procedure is given (Table 1), details depending on the oligo-dT/carrier combination should be obtained from the manufacturers protocols. First, total RNA is dissolved in a high salt buffer and heated briefly to 65-70 °C to disrupt secondary structures. Afterwards, annealing to oligo-dT is performed at room temperature. The oligo-dT molecules are linked to a carrier which allows washing off non-bound nucleic acids while retaining the oligo-dT bound poly(A)-RNA. After several washing steps under conditions less stringent than for annealing, poly(A)-RNA is eluted in water or low salt Tris buffer. Elution volumes vary from 20-250 µl for most kits accepting up to 1 mg of total RNA. A yield of 2-30 µg of poly(A)-RNA can be expected from 1 mg of total RNA, but might vary considerably with growth conditions of the mycelium from which the total RNA was extracted. mRNA should be stored in a −80 °C freezer. Concentration of eluted mRNA can be determined by OD measurement in slightly buffered solution like 10 mM Tris pH 7.0 (OD is pH dependent) using about 1/10 of the eluted volume. Quality of poly (A)-RNA can also be determined by Northern blot hybridization. As a rule of thumb, a transcript which gives a signal in a Northern blot with 20 µg of total RNA should be easily visible as a clear band without any downward smear on 0.1 µg of poly(A)-RNA. To determine size distribution of the eluted mRNAs, such a Northern blot can be probed with oligo-dT which should reveal a smear from ~0.2 to > 3kb. (Amounts less than 1 µg have to be checked by Northern hybridization as they cannot be seen on an agarose gel. To be visible on a gel, at least 1-2 µg poly(A)-RNA have to be used for electrophoresis, and what one usually sees are residual amounts of the ribosomal RNAs.) For some applications, mRNA has to be concentrated if the elution volume is too big. This can be done by ethanol precipitation, if more than ~800 µg of total RNA were used for mRNA preparation; with less starting material, mRNA amounts may be too low to result in a good recovery. Ethanol precipitation may be done with 0.1 volume 3 M sodium acetate (pH 5.2) and 2.5 volumes ethanol at −20 °C overnight. It should be noted that poly(A)-RNA preparations might still contain some residual genomic DNA. For many applications, this is not a problem; but for any application which involves a PCR step, it might be necessary to perform a DNase treatment before using the mRNA for downstream experiments. Several manufacturers offer DNases and buffer systems which can be used prior to a reverse transcription and PCR step without the need for buffer removal (e.g. DNaseI, amplification grade, from Invitrogen).

" Use of up to 1.5 mg total RNA from a phenol/chloroform extraction instead of 1 mg as starting material does improve yield. b The original protocol recommends 0.5x SSC for binding and 0.1x SSC for washing, but better results with Neurospora have been obtained with 1x and 0.2x SSC, respectively."

DNA isolation

DNA isolation extracts DNA from a cell in a pure form. First, the DNA is separated from cellular components such as proteins, RNA, and lipids. This is done by placing the chosen cells in a tube with a solution that mechanically, chemically, breaks the cells open. This solution contains enzymes, chemicals, and salts that breaks down the cells except for the DNA. It contains enzymes to dissolve proteins, chemicals to destroy all RNA present, and salts to help pull DNA out of the solution. Next, the DNA is separated from the solution by being spun in a centrifuge, which allows the DNA to collect in the bottom of the tube. After this cycle in the centrifuge the solution is poured off and the DNA is resuspended in a second solution that makes the DNA easy to work with in the future. This results in a concentrated DNA sample that contains thousands of copies of each gene. For large scale projects such as sequencing the human genome, all this work is done by robots.

The isolation of DNA from whole blood by a modified rapid method (RM) was tested using various detergents and buffer conditions. Extraction of DNA with either NP-40 or Triton X-100 gave a high yield of undegraded DNA in less than an hour. The concentration of magnesium ion in the buffers was critical to obtaining intact, high molecular weight (HMW) DNA. Greater than 10 mM $MgCl_2$ led to degradation. Addition of EDTA to the buffer inhibits this degradation. Preparation of DNA from blood stored at room temperature or incubated at 37°C for 24 hr resulted in the same amount and quality of DNA as from samples frozen at -70°C. DNA from blood samples that had undergone more than four freeze-thaw cycles was found to be partially degraded. The modified RM can be applied to extract DNA from as little as 10 μl of blood (340 ng of DNA) and from dried blood samples. DNA samples remained intact and undegraded for longer times when DNA was dissolved in higher concentrations of EDTA.

Gene therapy

A mutation in a gene can cause encoded proteins and the cells that rely on those proteins to malfunction. Conditions related to gene mutations are called genetic disorders. However, altering a patient's genes can sometimes be used to treat or cure a disease as well. Gene therapy can be used to replace a mutated gene with the correct copy of the gene, to inactivate or knockout the expression of a malfunctioning gene, or to introduce a foreign gene to the body to help fight disease.[6] Major diseases that can be treated with gene therapy include viral infections, cancers, and inherited disorders, including immune system disorders.

Gene therapy delivers a copy of the missing, mutated, or desired gene via a modified virus or vector to the patient's target cells so that a functional form of the protein can then be produced and incorporated into the body. These vectors are often siRNA. Treatment can be either in vivo or ex vivo. The therapy has to be repeated several times for the infected patient to continually be relieved, as repeated cell division and cell death slowly randomizes the body's ratio of functional-to-mutant genes. Gene therapy is an appealing alternative to some drug-based approaches, because gene therapy repairs the underlying genetic defect using the patients own cells with minimal side effects. Gene therapies are still in development and mostly used in research settings. All experiments and products are controlled by the U.S. FDA and the NIH. Classical gene therapies usually require efficient transfer of cloned genes into the disease cells so that the introduced genes are expressed at sufficiently high levels to change the patient's physiology. There are several different physicochemical and biological methods that can be used to transfer genes into human cells. The size of the DNA fragments that can be transferred is very limited, and often the transferred gene is not a conventional gene. Horizontal gene transfer is the transfer of genetic material from one cell to another that is not its offspring. Artificial horizontal gene transfer is a form of genetic engineering.

Heredity, Genes, and DNA

Perhaps the most fundamental property of all living things is the ability to reproduce. All organisms inherit the genetic information specifying their structure and function from their parents. Likewise, all cells arise from preexisting cells, so the genetic material must be replicated and passed from parent to progeny cell at each cell division. How genetic information is replicated and transmitted from cell to cell and organism to organism thus represents a question that is central to all of biology. Consequently, elucidation of the mechanisms of genetic transmission and identification of the genetic material as DNA were discoveries that formed the foundation of our current understanding of biology at the molecular level.

Genes and Chromosomes

The classical principles of genetics were deduced by Gregor Mendel in 1865, on the basis of the results of breeding experiments with peas. Mendel studied the inheritance of a number of well-defined traits, such as seed color, and was able to deduce general rules for their transmission. In all cases, he could correctly interpret the observed patterns of inheritance by assuming that each trait is determined by a pair of inherited factors, which are now called genes. One gene copy (called an allele) specifying each trait is inherited from each parent. For example, breeding two strains of peas—one having yellow seeds, and the other green seeds—yields the following results (Figure 3.1). The parental strains each have two identical copies of the gene specifying yellow (Y) or green (y) seeds, respectively. The progeny plants are therefore hybrids, having inherited one gene for yellow seeds (Y) and one for green seeds (y). All these progeny plants (the first filial, or F1, generation) have yellow seeds, so yellow (Y) is said to be dominant and green (y) recessive. The genotype (genetic composition) of the F1 peas is thus Yy, and their phenotype (physical appearance) is yellow. If one F1 offspring is bred with another, giving rise to F2 progeny, the genes for yellow and green seeds segregate in a characteristic manner such that the ratio between F2 plants with yellow seeds and those with green seeds.

Mendel's findings, apparently ahead of their time, were largely ignored until 1900, when Mendel's laws were rediscovered and their importance recognized. Shortly thereafter, the role of chromosomes as the carriers of genes was proposed. It was realized that most cells of higher plants and animals are diploid—containing two copies of each chromosome. Formation of the germ cells (the sperm and egg), however, involves a unique type of cell division (meiosis) in which only one member of each chromosome pair is transmitted to each progeny cell (Figure 3.2). Consequently, the sperm and egg are haploid, containing only one copy of each chromosome. The union of these two haploid cells at fertilization creates a new diploid organism, now containing one member of each chromosome pair derived from the male and one from the female parent. The behavior of chromosome pairs thus parallels that of genes, leading to the conclusion that genes are carried on chromosomes.

The fundamentals of mutation, genetic linkage, and the relationships between genes and chromosomes were largely established by experiments performed with the fruit fly, Drosophila melanogaster. Drosophila can be easily maintained in the laboratory, and they reproduce about every two weeks, which is a considerable advantage for genetic experiments. Indeed, these features continue to make Drosophila an organism of choice for genetic studies of animals, particularly the genetic analysis of development and differentiation.

In the early 1900s, a number of genetic alterations (mutations) were identified in Drosophila, usually affecting readily observable characteristics such as eye color or wing shape. Breeding experiments indicated that some of the genes governing these traits are inherited independently of each other, suggesting that these genes are located on different chromosomes that segregate independently during meiosis (Figure 3.3). Other genes, however, are frequently inherited together as paired characteristics. Such genes are said

Molecular Genetics

to be linked to each other by virtue of being located on the same chromosome. The number of groups of linked genes is the same as the number of chromosomes (four in Drosophila), supporting the idea that chromosomes are carriers of the genes.

Linkage between genes is not complete, however; chromosomes exchange material during meiosis, leading to recombination between linked genes. The frequency of recombination between two linked genes depends on the distance between them on the chromosome; genes that are close to each other recombine less frequently than do genes farther apart. Thus, the frequencies with which different genes recombine can be used to determine their relative positions on the chromosome, allowing the construction of genetic maps. By 1915, nearly a hundred genes had been defined and mapped onto the four chromosomes of Drosophila, leading to general acceptance of the chromosomal basis of heredity.

Genes and Enzymes

Early genetic studies focused on the identification and chromosomal localization of genes that control readily observable characteristics, such as the eye color of Drosophila. How these genes lead to the observed phenotypes, however, was unclear. The first insight into the relationship between genes and enzymes came in 1909, when it was realized that the inherited human disease phenylketonuria (see Molecular Medicine in Chapter 2) results from a genetic defect in metabolism of the amino acid phenylalanine. This defect was hypothesized to result from a deficiency in the enzyme needed to catalyze the relevant metabolic reaction, leading to the general suggestion that genes specify the synthesis of enzymes.

Clearer evidence linking genes with the synthesis of enzymes came from experiments of George Beadle and Edward Tatum, performed in 1941 with the fungus Neurospora crassa. In the laboratory, Neurospora can be grown on minimal or rich media similar to those discussed in Chapter 1 for the growth of E. coli. For Neurospora, minimal media consist only of salts, glucose, and biotin; rich media are supplemented with amino acids, vitamins, purines, and pyrimidines. Beadle and Tatum isolated mutants of Neurospora that grew normally on rich media but could not grow on minimal media. Each mutant was found to require a specific nutritional supplement, such as a particular amino acid, for growth. Furthermore, the requirement for a specific nutritional supplement correlated with the failure of the mutant to synthesize that particular compound. Thus, each mutation resulted in a deficiency in a specific metabolic pathway. Since such metabolic pathways were known to be governed by enzymes, the conclusion from these experiments was that each gene specified the structure of a single enzyme—

the one gene-one enzyme hypothesis. Many en-zymes are now known to consist of multiple polypeptides, so the currently accepted statement of this hypothesis is that each gene specifies the structure of a single polypeptide chain.

Identification of DNA as the Genetic Material

Understanding the chromosomal basis of heredity and the relationship between genes and enzymes did not in itself provide a molecular explanation of the gene. Chromosomes contain proteins as well as DNA, and it was initially thought that genes were proteins. The first evidence leading to the identification of DNA as the genetic material came from studies in bacteria. These experiments represent a prototype for current approaches to defining the function of genes by introducing new DNA sequences into cells, as discussed later in this chapter.

The experiments that defined the role of DNA were derived from studies of the bacterium that causes pneumonia (Pneumococcus). Virulent strains of Pneumococcus are surrounded by a polysaccharide capsule that protects the bacteria from attack by the immune system of the host. Because the capsule gives bacterial colonies a smooth appearance in culture, encapsulated strains are denoted S. Mutant strains that have lost the ability to make a capsule (denoted R) form rough-edged colonies in culture and are no longer lethal when inoculated into mice. In 1928 it was observed that mice inoculated with nonencapsulated (R) bacteria plus heat-killed encapsulated (S) bacteria developed pneumonia and died. Importantly, the bacteria that were then isolated from these mice were of the S type. Subsequent experiments showed that a cell-free extract of S bacteria was similarly capable of converting (or transforming) R bacteria to the S state. Thus, a substance in the S extract (called the transforming principle) was responsible for inducing the genetic transformation of R to S bacteria.

In 1944 Oswald Avery, Colin MacLeod, and Maclyn McCarty established that the transforming principle was DNA, both by purifying it from bacterial extracts and by demonstrating that the activity of the transforming principle is abolished by enzymatic digestion of DNA but not by digestion of proteins (Figure 3.6). Although these studies did not immediately lead to the acceptance of DNA as the genetic material, they were extended within a few years by experiments with bacterial viruses. In particular, it was shown that, when a bacterial virus infects a cell, the viral DNA rather than the viral protein must enter the cell in order for the virus to replicate. Moreover, the parental viral DNA (but not the protein) is transmitted to progeny virus particles. The concurrence of these results with continuing studies of the activity of

Molecular Genetics

DNA in bacterial transformation led to acceptance of the idea that DNA is the genetic material.

DNA as the agent of heredity

In 1869 Swiss chemist Johann Friedrich Miescher extracted a substance containing nitrogen and phosphorus from cell nuclei. The substance was originally called nuclein, but it is now known as deoxyribonucleic acid, or DNA. DNA is the chemical component of the chromosomes that is chiefly responsible for their staining properties in microscopic preparations. Since the chromosomes of eukaryotes contain a variety of proteins in addition to DNA, the question naturally arose whether the nucleic acids or the proteins, or both together, were the carriers of the genetic information. Until the early 1950s most biologists were inclined to believe that the proteins were the chief carriers of heredity. Nucleic acids contain only four different unitary building blocks, but proteins are made up of 20 different amino acids. Proteins therefore appeared to have a greater diversity of structure, and the diversity of the genes seemed at first likely to rest on the diversity of the proteins.

Evidence that DNA acts as the carrier of the genetic information was first firmly demonstrated by exquisitely simple microbiological studies. In 1928 English bacteriologist Frederick Griffith was studying two strains of the bacterium Streptococcus pneumoniae; one strain was lethal to mice (virulent) and the other was harmless (avirulent). Griffith found that mice inoculated with either the heat-killed virulent bacteria or the living avirulent bacteria remained free of infection, but mice inoculated with a mixture of both became infected and died. It seemed as if some chemical "transforming principle" had transferred from the dead virulent cells into the avirulent cells and changed them. In 1944 American bacteriologist Oswald T. Avery and his coworkers found that the transforming factor was DNA. Avery and his research team obtained mixtures from heat-killed virulent bacteria and inactivated either the proteins, polysaccharides (sugar subunits), lipids, DNA, or RNA (ribonucleic acid, a close chemical relative of DNA) and added each type of preparation individually to avirulent cells. The only molecular class whose inactivation prevented transformation to virulence was DNA. Therefore, it seemed that DNA, because it could transform, must be the hereditary material.

A similar conclusion was reached from the study of bacteriophages, viruses that attack and kill bacterial cells. From a host cell infected by one bacteriophage, hundreds of bacteriophage progeny are produced. In 1952 American biologists Alfred D. Hershey and Martha Chase prepared two populations of bacteriophage particles. In one population, the outer protein coat of the bacteriophage was labeled with a radioactive isotope; in the other, the DNA was labeled. After allowing both populations to attack bacteria, Hershey and Chase found that only when DNA was labeled did the progeny bacteriophage contain radioactivity. Therefore, they concluded that DNA is injected into the bacterial cell, where it directs the synthesis of numerous complete bacteriophages at the expense of the host. In other words, in bacteriophages DNA is the hereditary material responsible for the fundamental characteristics of the virus.

Today the genetic makeup of most organisms can be transformed using externally applied DNA, in a manner similar to that used by Avery for bacteria. Transforming DNA is able to pass through cellular and nuclear membranes and then integrate into the chromosomal DNA of the recipient cell. Furthermore, using modern DNA technology, it is possible to isolate the section of chromosomal DNA that constitutes an individual gene, manipulate its structure, and reintroduce it into a cell to cause changes that show beyond doubt that the DNA is responsible for a large part of the overall characteristics of an organism. For reasons such as these, it is now accepted that, in all living organisms, with the exception of some viruses, genes are composed of DNA.

An analysis of concepts in practice, the classical gene and molecular gene concepts

It has been argued, against gene skepticism, that biologists have a coherent, precise, and uniform way to conceive of genes at the molecular level. The analysis underlying this argument begins by distinguishing between two different ways contemporary geneticists think about genes. Classical geneticists often conceived of genes as the functional units in chromosomes, differences in which cause differences in phenotypes. Today, in contexts where genes are identified by way of observed phenotypic differences, geneticists still conceive of genes in this classical way, as the functional units in DNA whose differences are causing the observed differences in phenotypes. This way of conceiving of genes is called the classical gene concept (Waters 1994). But contemporary geneticists also think about genes in a different way by invoking a molecular-level concept. The molecular gene concept stems from the idea that genes are units in DNA that function to determine linear sequences in molecules synthesized via DNA expression. According to this analysis, both concepts are at work in contemporary geneticists. Moss 2003 also distinguishes between two contemporary gene concepts, which he calls "genes-P (preformationist)" and "genes-D (developmental)". He argues that conflation of these concepts leads to erroneous thinking in genetics.

Much confusion concerning the classical way to think about genes is due to the fact that geneticists have sometimes talked as if classically conceived

Molecular Genetics

genes are for gross phenotypic characters (phenotypes) or as if individual genes produce phenotypes. This talk was very misleading on the part of classical geneticists and continues to be misleading in the context of contemporary genetics. The production of a gross phenotypic character, such as purple eye-color, involves all sorts of genetic and extra-genetic factors including various cellular enzymes and structures, tissue arrangements, and environmental factors. In addition, it is not clear what, if any, gross phenotypic level functions can be attributed to individual genes. For example, it is no clearer today than it was in Morgan's day that the function of the purple gene discussed in section 2.1 is to contribute to the production of eye color. Mutations in this gene affect a number of gross phenotypic level traits. Legitimate explanatory reasoning invoking the classical gene concept does not depend on any baggage concerning what genes are for or what function a gene might have in development. What the explanatory reasoning depends on is the difference principle, that is, the principle that some difference in the gene causes certain phenotypic differences in particular genetic and environmental contexts (section 2.1). Many gene-based explanations in contemporary biology are best understood in terms of the classical gene concept and the difference principle.

Perhaps the reason gene skeptics overlooked the molecular gene concept is that they were searching for the wrong kind of concept. The concept is not a purely physicochemical concept, and it does not provide a single partition of DNA into separate genes. Instead, it is a functional concept that provides a uniform way to think about genes that can be applied to pick out different DNA segments in different investigative or explanatory contexts. The basic molecular concept, according to this analysis, is the concept of a gene for a linear sequence in a product of DNA expression:

A gene g for linear sequence l in product p synthesized in cellular context c is a potentially replicating nucleotide sequence, n, usually contained in DNA, that determines the linear sequence l in product p at some stage of DNA expression (Waters 2000)

The concept of the molecular gene can be presented as a 4-tuple: <n,l,p,c>. This analysis shows how geneticists can consistently include introns as part of a gene in one epistemic context and not in another. If the context involves identifying a gene for a primary, preprocessed RNA molecule, then the gene includes the introns as well as the exons. If the context involves identifying the gene for the resulting polypeptide, then the gene includes only the exons. Hence, in the case of DNA expression that eventually leads to the synthesis of a given polypeptide, geneticists might talk as if "the" gene included the intron (in which case they would be referring to the gene for the primary, preprocessed RNA) and yet also talk as if "the" gene excluded the introns (in which case they would be referring to the gene for the mature RNA or polypeptide). Application of the molecular gene concept is not ambiguous; in fact, it is remarkably precise provided one specifies the values for the variables in the expression "gene for linear sequence l in product p synthesized in cellular context c."

Gene skeptics have suggested that there is a lack of coherence in gene talk because biologists often talk as if genes code for polypeptides, but then turn around and talk about genes for RNA molecules that are not translated into polypeptides (including genes for RNA [tRNA], ribosomal RNA [rRNA], and interference RNA [iRNA]). This account shows that conceiving of genes for rRNA involves the same idea as conceiving of genes for polypeptides. In both cases, the gene is the segment of DNA, split or not, that determines the linear sequence in the molecule of interest.

An advantage of this analysis is that it emphasizes the limitations of gene-centered explanations while clarifying the distinctive causal role genes play in the syntheses of RNA and polypeptides: genes determine the linear sequences of primary RNA transcripts and often play a distinctive role, though not exclusive, in determining the sequence of amino acids in polypeptides.

Analysis of reference in practice, how Molecularization changed reference

Weber (2005) examines the evolution of the gene concept by tracing changes in the reference of the term gene through the history of genetics. The reference or extension of a term is the set of objects to which it reference. Weber adopts a mixed theory of refence. According to mixed theories, the reference of a term is determined how the relevant linguistic community causally interacts with potential referents as well as how they describe potential referents. This theory leads Weber to pay close attention, not just to how geneticists theorized about genes or used the concept to explain phenomena, but also how they conducted their laboratory investigations. Following Kitcher (1978, 1982), he examines ways in which modes of reference changed over time.

Weber identifies six different gene concepts, beginning with Darwin's pangene concept (1868) and ending with the contemporary concept of molecular genetics. He distinguishes the contemporary molecular concept from the classical (or 'neoclassical') one on the basis of how geneticists described their functional role (RNA/protein coding versus general function unit), their material basis (RNA/DNA versus chromosome), and their structure (discontinuous linear -- with introns and exons versus continuous linear)

Molecular Genetics

as well as on the basis of the criteria experimentalists used to identify genes (by gene product versus complementation test).

Weber examines how the investigation of several particular Drosophila genes changed as the science of genetics developed. His study shows that the methods of molecular genetics provided new ways to identify genes that were first identified by classical techniques. The reference of the term changed, not simply as a result of theoretical developments, but also as a result of the implementation of new methods to identify genes. He concludes that unlike concepts of physical science that have been analyzed by philosophers, the gene concept has a "nonessentialistic character that allows biologists to lay down different natural classifications, depending on the investigative methods available as well as on theoretical interests" (Weber 2005, p. 228). Weber calls this feature "floating references".

Recombination and genetic linkage

The diploid nature of chromosomes allows for genes on different chromosomes to assort independently or be separated from their homologous pair during sexual reproduction wherein haploid gametes are formed. In this way new combinations of genes can occur in the offspring of a mating pair. Genes on the same chromosome would theoretically never recombine. However, they do, via the cellular process of chromosomal crossover. During crossover, chromosomes exchange stretches of DNA, effectively shuffling the gene alleles between the chromosomes. This process of chromosomal crossover generally occurs during meiosis, a series of cell divisions that creates haploid cells.

The first cytological demonstration of crossing over was performed by Harriet Creighton and Barbara McClintock in 1931. Their research and experiments on corn provided cytological evidence for the genetic theory that linked genes on paired chromosomes do in fact exchange places from one homolog to the other.

The probability of chromosomal crossover occurring between two given points on the chromosome is related to the distance between the points. For an arbitrarily long distance, the probability of crossover is high enough that the inheritance of the genes is effectively uncorrelated.[56] For genes that are closer together, however, the lower probability of crossover means that the genes demonstrate genetic linkage; alleles for the two genes tend to be inherited together. The amounts of linkage between a series of genes can be combined to form a linear linkage map that roughly describes the arrangement of the genes along the chromosome.

Genetic code

The genetic code is a set of rules defining how the four-letter code of DNA is translated into the 20-letter code of amino acids, which are the building blocks of proteins. The genetic code is a set of three-letter combinations of nucleotides called codons, each of which corresponds to a specific amino acid or stop signal. The concept of codons was first described by Francis Crick and his colleagues in 1961. During the same year, Marshall Nirenberg and Heinrich Matthaei performed experiments that began deciphering the genetic code. They showed that the RNA sequence UUU specifically coded for the amino acid phenylalanine. Following this discovery, Nirenberg, Philip Leder, and Gobind Khorana identified the rest of the genetic code and fully described each three-letter codon and its corresponding amino acid.

There are 64 possible permutations, or combinations, of three-letter nucleotide sequences that can be made from the four nucleotides. Of these 64 codons, 61 represent amino acids, and three are stop signals. Although each codon is specific for only one amino acid (or one stop signal), the genetic code is described as degenerate, or redundant, because a single amino acid may be coded for by more than one codon. It is also important to note that the genetic code does not overlap, meaning that each nucleotide is part of only one codon-a single nucleotide cannot be part of two adjacent codons. Furthermore, the genetic code is nearly universal, with only rare variations reported. For instance, mitochondria have an alternative genetic code with slight variations.

Genes generally express their functional effect through the production of proteins, which are complex molecules responsible for most functions in the cell. Proteins are made up of one or more polypeptide chains, each of which is composed of a sequence of amino acids, and the DNA sequence of a gene (through an RNA intermediate) is used to produce a specific amino acid sequence. This process begins with the production of an RNA molecule with a sequence matching the gene's DNA sequence, a process called transcription.

This messenger RNA molecule is then used to produce a corresponding amino acid sequence through a process called translation. Each group of three nucleotides in the sequence, called a codon, corresponds either to one of the twenty possible amino acids in a protein or an instruction to end the amino acid sequence; this correspondence is called the genetic code. The flow of information is unidirectional: information is transferred from nucleotide sequences into the amino acid sequence of proteins, but it never transfers from protein back into the sequence of DNA—a phenomenon Francis Crick called the central dogma of molecular biology.

Molecular Genetics

The specific sequence of amino acids results in a unique three-dimensional structure for that protein, and the three-dimensional structures of proteins are related to their functions. Some are simple structural molecules, like the fibers formed by the protein collagen. Proteins can bind to other proteins and simple molecules, sometimes acting as enzymes by facilitating chemical reactions within the bound molecules (without changing the structure of the protein itself). Protein structure is dynamic; the protein hemoglobin bends into slightly different forms as it facilitates the capture, transport, and release of oxygen molecules within mammalian blood.

A single nucleotide difference within DNA can cause a change in the amino acid sequence of a protein. Because protein structures are the result of their amino acid sequences, some changes can dramatically change the properties of a p rotein by destabilizing the structure or changing the surface of the protein in a way that changes its interaction with other proteins and molecules. For example, sickle-cell anemia is a human genetic disease that results from a single base difference within the coding region for the ß-globin section of hemoglobin, causing a single amino acid change that changes hemoglobin's physical properties. Sickle-cell versions of hemoglobin stick to themselves, stacking to form fibers that distort the shape of red blood cells carrying the protein. These sickle-shaped cells no longer flow smoothly through blood vessels, having a tendency to clog or degrade, causing the medical problems associated with this disease.

Some DNA sequences are transcribed into RNA but are not translated into protein products—such RNA molecules are called non-coding RNA. In some cases, these products fold into structures which are involved in critical cell functions (e.g. ribosomal RNA and transfer RNA). RNA can also have regulatory effects through hybridization interactions with other RNA molecules (e.g. microRNA).

3

Molecular Biology

INTRODUCTION

The basic concept of molecular biology is to investigate the activities of an organism's makeup at a sub-cellular level. The focus is on the sequence of the DNA (the genes), the rate and timing of the expression of those genes, the mechanisms involved in expressing those genes, and the effect that they have on the whole cell and ultimately the whole organism. Nucleic acids and Proteins are the working entities of molecular mechanisms. Genes are made of nucleic acids, usually DNA. Some simple genetic systems such as viruses have RNA genes.

Recombinant DNA techniques have increased our understanding of genetic disorders and facilitated their diagnosis and may eventually lead to therapy. Expression may be assessed and quantified by Northern blot analysis, whereas in situ hybridization may localize the transcript. Southern blotting has been the standard method to identify gene mutations and RFLPs, which are useful in the diagnosis of many different disorders. Special multiallelic RFLPs produce a DNA fingerprint particularly useful in paternity testing and forensic medicine. PCR has greatly reduced the time and expense to diagnose genetic disorders, and has an enormous number of applications both clinically and at the research level. These techniques along with DNA sequencing are currently being used in the human genome project, which may some day allow "DNA typing" of individuals for any gene of interest. There is no doubt that new techniques in molecular biology will continue to evolve so that the goal of gene therapy for many disorders may be possible in the future.

"...not so much a technique as an approach, an approach from the viewpoint of the so-called basic sciences with the leading idea of searching below the large-scale manifestations of classical biology for the corresponding molecular plan. It is concerned particularly with the forms of biological

Molecular Biology

molecules and [...] is predominantly three-dimensional and structural—which does not mean, however, that it is merely a refinement of morphology. It must at the same time inquire into genesis and function."

Actually molecular biology is the branch of biology that studies the structure and activity of macromolecules essential to life (and especially with their genetic role) cistron, gene, factor - (genetics) a segment of DNA that is involved in producing a polypeptide chain; it can include regions preceding and following the coding DNA as well as introns between the exons; it is considered a unit of heredity; "genes were formerly called factors" biological science, biology the science that studies living organisms biotech, biotechnology - the branch of molecular biology that studies the use of microorganisms to perform specific industrial processes; "biotechnology produced genetically altered bacteria that solved the problem"

Relationship to other biological sciences

Researchers in molecular biology use specific techniques native to molecular biology but increasingly combine these with techniques and ideas from genetics and biochemistry. There is not a defined line between these disciplines. The figure to the right is a schematic that depicts one possible view of the relationships between the fields:

- Biochemistry is the study of the chemical substances and vital processes occurring in live organisms. Biochemists focus heavily on the role, function, and structure of biomolecules. The study of the chemistry behind biological processes and the synthesis of biologically active molecules are examples of biochemistry.
- Genetics is the study of the effect of genetic differences on organisms. This can often be inferred by the absence of a normal component (e.g. one gene). The study of "mutants" – organisms which lack one or more functional components with respect to the so-called "wild type" or normal phenotype. Genetic interactions (epistasis) can often confound simple interpretations of such "knockout" studies.
- Molecular biology is the study of molecular underpinnings of the processes of replication, transcription, translation, and cell function. The central dogma of molecular biology where genetic material is transcribed into RNA and then translated into protein, despite being an oversimplified picture of molecular biology, still provides a good starting point for understanding the field. This picture, however, is undergoing revision in light of emerging novel roles for RNA.

Much of molecular biology is quantitative, and recently much work has been done at its interface with computer science in bioinformatics and computational biology. In the early 2000s, the study of gene structure and function, molecular genetics, has been among the most prominent sub-fields of molecular biology. Increasingly many other areas of biology focus on molecules, either directly studying interactions in their own right such as in cell biology and developmental biology, or indirectly, where molecular techniques are used to infer historical attributes of populations or species, as in fields in evolutionary biology such as population genetics and phylogenetics. There is also a long tradition of studying biomolecules "from the ground up" in biophysics.

Importance of Protein in Molecular Biology

Members of the HMGA (a.k.a. HMGI/Y) family of 'high mobility group' (HMG) proteins participate in a wide variety of nuclear processes ranging from chromosome and chromatin mechanics to acting as architectural transcription factors that regulate the expression of numerous genes in vivo. As a consequence, they function in the cell as highly connected 'nodes' of protein–DNA and protein–protein interactions that influence a diverse array of normal biological processes including growth, proliferation, differentiation and death. The HMGA proteins, likewise, participate in pathological processes by, for example, acting as regulators of viral gene transcription and by serving as host-supplied proteins that facilitate retroviral integration. HMGA genes are bona fide proto-oncogenes that promote tumor progression and metastasis when overexpressed in cells. High constitutive HMGA protein levels are among the most consistent feature observed in all types of cancers with increasing concentrations being correlated with increasing malignancy. The intrinsic attributes that endow the HMGA proteins with these remarkable abilities are a combination of structural, biochemical and biological characteristics that are unique to these proteins. HMGA proteins have little, if any, secondary structure while free in solution but undergo disordered-to-ordered structural transitions when bound to substrates such as DNA or other proteins. Each protein contains three copies of a conserved DNA-binding peptide motif called the 'AT-hook' that preferentially binds to the minor groove of stretches of AT-rich sequence. In vivo HMGA proteins specifically interact with a large number of other proteins, most of which are transcription factors. They are also subject to many types of in vivo biochemical modifications that markedly influence their ability to interact with DNA substrates, other proteins and chromatin. And, most importantly, both the transcription of HMGA genes and the biochemical modifications of HMGA proteins are direct downstream targets of numerous signal transduction pathways making them exquisitely responsive to various environmental influences. This review covers recent advances that have contributed to our understanding of how this constellation of structural and

Molecular Biology

biological features allows the HMGA proteins to serve as central 'hubs' of nuclear function.

What is a protein in a cell?

They do most of the work in cells and are required for the structure, function, and regulation of the body's tissues and organs. Proteins are made up of hundreds or thousands of smaller units called amino acids, which are attached to one another in long chains.

What are proteins and what do they do?

Proteins are large, complex molecules that play many critical roles in the body. They do most of the work in cells and are required for the structure, function, and regulation of the body's tissues and organs.

Proteins are made up of hundreds or thousands of smaller units called amino acids, which are attached to one another in long chains. There are 20 different types of amino acids that can be combined to make a protein. The sequence of amino acids determines each protein's unique 3-dimensional structure and its specific function.

Proteins can be described according to their large range of functions in the body, listed in alphabetical order:

Examples of protein functions

Function	Description	Example
Antibody	Antibodies bind to specific foreign particles, such as viruses and bacteria, to help protect the body.	Immunoglobulin (IgG) (illustration)
Enzyme	Enzymes carry out almost all of the thousands of chemical reactions that take place in cells. They also assist with the formation of new molecules by reading the genetic information stored in DNA.	Phenylalanine hydroxylase (illustration)
Messenger	Messenger proteins, such ass some types of hormones, transmit signals to coordinate biological processes between different cells, tissues, and organs.	Growth hormone (illustration)
Structural component	These proteins provide structure and support for cells. On a larger scale, they also allow the body to move.	Actin (illustration)
Transport/ storage	These proteins bind and carry atoms and small molecules within cells and throughout the body.	Ferritin (illustration)

Function of Protein

We have seen that each type of protein consists of a precise sequence of amino acids that allows it to fold up into a particular three-dimensional shape, or conformation. But proteins are not rigid lumps of material. They can have precisely engineered moving parts whose mechanical actions are coupled to chemical events. It is this coupling of chemistry and movement that gives proteins the extraordinary capabilities that underlie the dynamic processes in living cells.

In this section, we explain how proteins bind to other selected molecules and how their activity depends on such binding. We show that the ability to bind to other molecules enables proteins to act as catalysts, signal receptors, switches, motors, or tiny pumps. The examples we discuss in this chapter by no means exhaust the vast functional repertoire of proteins. However, the specialized functions of many of the proteins you will encounter elsewhere in this book are based on similar principles.

- Antibodies - are specialized proteins involved in defending the body from antigens (foreign invaders). They can travel through the bloodstream and are utilized by the immune system to identify and defend against bacteria, viruses, and other foreign intruders. One way antibodies counteract antigens is by immobilizing them so that they can be destroyed by white blood cells.
- Contractile Proteins - are responsible for movement. Examples include actin and myosin. These proteins are involved in muscle contraction and movement.
- Enzymes - are proteins that facilitate biochemical reactions. They are often referred to as catalysts because they speed up chemical reactions. Examples include the enzymes lactase and pepsin. Lactase breaks down the sugar lactose found in milk. Pepsin is a digestive enzyme that works in the stomach to break down proteins in food.
- Hormonal Proteins - are messenger proteins which help to coordinate certain bodily activities. Examples include insulin, oxytocin, and somatotropin. Insulin regulates glucose metabolism by controlling the blood-sugar concentration. Oxytocin stimulates contractions in females during childbirth. Somatotropin is a growth hormone that stimulates protein production in muscle cells.
- Structural Proteins - are fibrous and stringy and provide support. Examples include keratin, collagen, and elastin. Keratins strengthen protective coverings such as skin, hair, quills, feathers, horns, and beaks. Collagens and elastin provide support for connective tissues such as tendons and ligaments.

Molecular Biology

- Storage Proteins - store amino acids. Examples include ovalbumin, casein, ferritin. Ovalbumin is found in egg whites and casein is a milk-based protein. Ferritin stores iron in hemoglobin.
- Transport Proteins - are carrier proteins which move molecules from one place to another around the body. Examples include hemoglobin and cytochromes. Hemoglobin transports oxygen through the blood via red blood cells. Cytochromes operate in the electron transport chain as electron carrier proteins.

STRUCTURE OF PROTEIN

There are four levels of protein structure. These levels are distinguished from one another by the degree of complexity in the polypeptide chain (linked amino acids). The four levels of protein structure are primary, secondary, tertiary, and quaternary structure. A single protein molecule may contain one or more of these protein structure types. The structure of a protein determines its function.

For example, collagen has a super-coiled helical shape. It is long, stringy, strong, and resembles a rope. This structure is great for providing support. Hemoglobin, on the other hand, is a globular protein that is folded and compact. Its spherical shape is useful for maneuvering through blood vessels.

The four levels of protein structure are distinguished from one another by the degree of complexity in the polypeptide chain. A single protein molecule may contain one or more of the protein structure types.

- Primary Structure - describes the unique order in which amino acids are linked together to form a protein. Proteins are constructed from a set of 20 amino acids. Generally, amino acids have the following structural properties:
- A carbon (the alpha carbon) bonded to the four groups below:
- A hydrogen atom (H)
- A Carboxyl group (-COOH)
- An Amino group ($-NH_2$)
- A "variable" group or "R" group

All amino acids have the alpha carbon bonded to a hydrogen atom, carboxyl group, and amino group. The "R" group varies among amino acids and determines the differences between these protein monomers. The amino acid sequence of a protein is determined by the information found in the cellular genetic code. The order of amino acids in a polypeptide chain is unique and specific to a particular protein. Altering a single amino acid causes a gene mutation, which most often results in a non-functioning protein.

- Secondary Structure - refers to the coiling or folding of a polypeptide chain that gives the protein its 3-D shape. There are two types of secondary structures observed in proteins. One type is the alpha (a) helix structure. This structure resembles a coiled spring and is secured by hydrogen bonding in the polypeptide chain. The second type of secondary structure in proteins is the beta (ß) pleated sheet. This structure appears to be folded or pleated and is held together by hydrogen bonding between polypeptide units of the folded chain that lie adjacent to one another.
- Tertiary Structure - refers to the comprehensive 3-D structure of the polypeptide chain of a protein. There are several types of bonds and forces that hold a protein in its tertiary structure. Hydrophobic interactions greatly contribute to the folding and shaping of a protein. The "R" group of the amino acid is either hydrophobic or hydrophilic. The amino acids with hydrophilic "R" groups will seek contact with their aqueous environment, while amino acids with hydrophobic "R" groups will seek to avoid water and position themselves towards the center of the protein. Hydrogen bonding in the polypeptide chain and between amino acid "R" groups helps to stabilize protein structure by holding the protein in the shape established by the hydrophobic interactions. Due to protein folding, ionic bonding can occur between the positively and negatively charged "R" groups that come in close contact with one another. Folding can also result in covalent bonding between the "R" groups of cysteine amino acids. This type of bonding forms what is called a disulfide bridge. Interactions called van der Waals forces also assist in the stabilization of protein structure. These interactions pertain to the attractive and repulsive forces that occur between molecules that become polarized. These forces contribute to the bonding that occurs between molecules.
- Quaternary Structure - refers to the structure of a protein macromolecule formed by interactions between multiple polypeptide chains. Each polypeptide chain is referred to as a subunit. Proteins with quaternary structure may consist of more than one of the same type of protein subunit. They may also be composed of different subunits. Hemoglobin is an example of a protein with quaternary structure. Hemoglobin, found in the blood, is an iron-containing protein that binds oxygen molecules. It contains four subunits: two alpha subunits and two beta subunits.

Molecular Biology

PROTEIN STRUCTURE DETERMINATION

The three-dimensional shape of a protein is determined by its primary structure. The order of amino acids establishes a protein's structure and specific function. The distinct instructions for the order of amino acids are designated by the genes in a cell. When a cell perceives a need for protein synthesis, the DNA unravels and is transcribed into an RNA copy of the genetic code. This process is called DNA transcription. The RNA copy is then translated to produce a protein. The genetic information in the DNA determines the specific sequence of amino acids and the specific protein that is produced. Proteins are examples of one type of biological polymer. Along with proteins, carbohydrates, lipids, and nucleic acids constitute the four major classes of organic compounds in living cells.

How do genes direct the production of proteins?

Most genes contain the information needed to make functional molecules called proteins. (A few genes produce other molecules that help the cell assemble proteins.) The journey from gene to protein is complex and tightly controlled within each cell. It consists of two major steps: transcription and translation. Together, transcription and translation are known as gene expression.

During the process of transcription, the information stored in a gene's DNA is transferred to a similar molecule called RNA (ribonucleic acid) in the cell nucleus. Both RNA and DNA are made up of a chain of nucleotide bases, but they have slightly different chemical properties. The type of RNA that contains the information for making a protein is called messenger RNA (mRNA) because it carries the information, or message, from the DNA out of the nucleus into the cytoplasm.

Translation, the second step in getting from a gene to a protein, takes place in the cytoplasm. The mRNA interacts with a specialized complex called a ribosome, which "reads" the sequence of mRNA bases. Each sequence of three bases, called a codon, usually codes for one particular amino acid. (Amino acids are the building blocks of proteins.) A type of RNA called transfer RNA (tRNA) assembles the protein, one amino acid at a time. Protein assembly continues until the ribosome encounters a "stop" codon (a sequence of three bases that does not code for an amino acid).

The flow of information from DNA to RNA to proteins is one of the fundamental principles of molecular biology. It is so important that it is sometimes called the "central dogma."

The Human Genome

he order of building blocks in a strand of DNA makes up a "sequence." We can read a DNA sequence like letters in a book. In fact, we know the sequence of the entire human genome—all 3 billion letters. That's enough information to fill roughly 1,000 200-page books!

Contained within the 3 billion letters of the human genome are about 21,000 genes. Most of our known genes code for proteins, but some code for RNA molecules.

DNA; The Same Chemical

The stringy stuff in the test tube is DNA. But you can't tell which one of these organisms it came from just by looking at it. That's because DNA looks exactly the same in every organism on Earth.

All living things have DNA. And whether it comes from you, a pea plant, or your pet rat, it's all the same molecule. It's the order of the letters in the code that makes each organism different.

DNA makes you unique

All humans have the same genes arranged in the same order. And more than 99.9% of our DNA sequence is the same. But the few differences between us (all 1.4 million of them!) are enough to make each one of us unique. On average, a human gene will have 1-3 bases that differ from person to person. These differences can change the shape and function of a protein, or they can change how much protein is made, when it's made, or where it's made.

All Proteins Bind to Other Molecules

The biological properties of a protein molecule depend on its physical interaction with other molecules. Thus, antibodies attach to viruses or bacteria to mark them for destruction, the enzyme hexokinase binds glucose and ATP so as to catalyze a reaction between them, actin molecules bind to each other to assemble into actin filaments, and so on. Indeed, all proteins stick, or bind, to other molecules. In some cases, this binding is very tight; in others, it is weak and short-lived. But the binding always shows great specificity, in the sense that each protein molecule can usually bind just one or a few molecules out of the many thousands of different types it encounters. The substance that is bound by the protein—no matter whether it is an ion, a small molecule, or a macromolecule— is referred to as a ligand for that protein (from the Latin word ligare, meaning "to bind").

The ability of a protein to bind selectively and with high affinity to a ligand depends on the formation of a set of weak, noncovalent bonds—hydrogen bonds, ionic bonds, and van der Waals attractions—plus favorable hydrophobic interaction. Because each individual bond is weak, an effective binding interaction requires that many weak bonds be formed simultaneously. This is possible only if the surface contours of the ligand molecule fit very closely to the protein, matching it like a hand in a glove.

The region of a protein that associates with a ligand, known as the ligand's binding site, usually consists of a cavity in the protein surface formed by a particular arrangement of amino acids. These amino acids can belong to different portions of the polypeptide chain that are brought together when the protein folds (Figure 3-38). Separate regions of the protein surface generally provide binding sites for different ligands, allowing the protein's activity to be regulated, as we shall see later. And other parts of the protein can serve as a handle to place the protein in a particular location in the cell—an example is the SH2 domain discussed previously, which is often used to move a protein containing it to sites in the plasma membrane in response to particular signals.

Although the atoms buried in the interior of the protein have no direct contact with the ligand, they provide an essential scaffold that gives the surface its contours and chemical properties. Even small changes to the amino acids in the interior of a protein molecule can change its three-dimensional shape enough to destroy a binding site on the surface.

The Details of a Protein's Conformation Determine Its Chemistry

Proteins have impressive chemical capabilities because the neighboring chemical groups on their surface often interact in ways that enhance the chemical reactivity of amino acid side chains. These interactions fall into two main categories.

First, neighboring parts of the polypeptide chain may interact in a way that restricts the access of water molecules to a ligand binding site. Because water molecules tend to form hydrogen bonds, they can compete with ligands for sites on the protein surface. The tightness of hydrogen bonds (and ionic interactions) between proteins and their ligands is therefore greatly increased if water molecules are excluded. Initially, it is hard to imagine a mechanism that would exclude a molecule as small as water from a protein surface without affecting the access of the ligand itself. Because of the strong tendency of water molecules to form water–water hydrogen bonds, however, water molecules exist in a large hydrogen-bonded network. In effect, a ligand binding site can be kept dry because it is energetically unfavorable for individual water molecules to break away from this network, as they must do to reach into a crevice on a protein's surface.

Second, the clustering of neighboring polar amino acid side chains can alter their reactivity. If a number of negatively charged side chains are forced together against their mutual repulsion by the way the protein folds, for example, the affinity of the site for a positively charged ion is greatly increased. In addition, when amino acid side chains interact with one another through hydrogen bonds, normally unreactive side groups (such as the –CH2OH on the serine) can become reactive, enabling them to enter into reactions that make or break selected covalent bonds.

The surface of each protein molecule therefore has a unique chemical reactivity that depends not only on which amino acid side chains are exposed, but also on their exact orientation relative to one another. For this reason, even two slightly different conformations of the same protein molecule may differ greatly in their chemistry

Protein Synthesis

When you picture protein, you might be thinking of elite body builders with their protein shakes, egg whites and plain chicken. It's true, all of these things contain protein. But when we really come down to it, proteins are tiny molecules inside cells, and they're required for all structure and function inside cells. Without them, our cells couldn't do their jobs and we would die. Like the furniture in your house, proteins wear out over time, so our cells are continuously making new proteins through the process of protein synthesis. Protein synthesis has two main steps, transcription and translation. Let's look at each step in detail next.

Transcription

DNA is the ultimate blueprint for the cell and holds all the instructions for making proteins. This molecule is so important that the cell wants to make copies of it to use in protein synthesis instead of moving the actual DNA around the cell. Think of DNA like the master copy of a document, like your birth certificate. To get other forms of identification you present a copy of your birth certificate, but not the real thing.

The copy of DNA is called mRNA (messenger RNA), another molecule that holds information in the cell. The process of transcription copies DNA to mRNA using a protein called RNA polymerase. After the copying, the mRNA is sent to a compartment of the cell called a ribosome, which does the next step, translation.

Translation

Ribosomes are compartments of the cell required for protein synthesis. During translation, they read the mRNA and tell another molecule called tRNA (transfer RNA) to get the building blocks for proteins, or amino acids. The ribosome strings together the amino acids according to the instructions

Molecular Biology

in the mRNA. After all the amino acids are put together, the protein folds up into a functional shape.

Proteins, like we mentioned, are responsible for all structure and function inside cells. All cells need proteins, not just our cells. So next, let's look at some key examples in different types of organisms.

Protein Synthesis in Animals

Humans are a type of animal, so our cells are considered animal cells. To understand how important proteins are in our cells, let's look at an example of what happens if there is a problem with protein synthesis. A protein called Rb (retinoblastoma protein) is responsible for controlling how often our cells divide, specifically in cells in the retina, an important part of our eyes. If cells divide out of control, we get cancer. Some people have a mutation, or a change in the DNA, in the instructions for making Rb. These people don't have a functional Rb protein, so their cells grow out of control and they get retinoblastoma, or cancer of the eye.

Protein Synthesis in Plants

Plants are a crucial part of our ecosystem as all other organisms depend on plants to get energy. Plants use a process called photosynthesis to turn carbon dioxide, water and sunlight into oxygen and a sugar called glucose. This happens through a series of chemical reactions that are made possible by proteins.

A particular protein called RuBisCo (Ribulose-1,5-bisphosphate carboxylase/oxygenase) incorporates carbon dioxide from the air into molecules inside the cell, which will ultimately make glucose.

Without glucose, humans and all other organisms that can't make their own food would die. When plants die off we no longer have oxygen either, which is also a product of photosynthesis. Clearly, although beyond microscopic, RuBisCo is essential for all life on Earth.

Proteins Bind to Other Proteins Through Several Types of Interfaces

Proteins can bind to other proteins in at least three ways. In many cases, a portion of the surface of one protein contacts an extended loop of polypeptide chain (a "string") on a second protein. Such a surface–string interaction, for example, allows the SH2 domain to recognize a phosphorylated polypeptide as a loop on a second protein, as just described, and it also enables a protein kinase to recognize the proteins that it will phosphorylate. A second type of protein–protein interface is formed when two a helices, one from each protein, pair together to form a coiled-coil. This type of protein interface is found in several families of gene regulatory proteins.

The most common way for proteins to interact, however, is by the precise matching of one rigid surface with that of another. Such interactions can be very tight, since a large number of weak bonds can form between two surfaces that match well. For the same reason, such surface–surface interactions can be extremely specific, enabling a protein to select just one partner from the many thousands of different proteins found in a cell.

Six Primary Functions of Proteins

Protein is an important substance found in every cell in the human body. In fact, except for water, protein is the most abundant substance in your body. This protein is manufactured by your body utilizing the dietary protein you consume. It is used in many vital processes and thus needs to be consistently replaced. You can accomplish this by regularly consuming foods that contain protein.

Repair and Maintenance

Protein is termed the building block of the body. It is called this because protein is vital in the maintenance of body tissue, including development and repair. Hair, skin, eyes, muscles and organs are all made from protein. This is why children need more protein per pound of body weight than adults; they are growing and developing new protein tissue.

Energy

Protein is a major source of energy. If you consume more protein than you need for body tissue maintenance and other necessary functions, your body will use it for energy. If it is not needed due to sufficient intake of other energy sources such as carbohydrates, the protein will be used to create fat and becomes part of fat cells.

Hormones

Protein is involved in the creation of some hormones. These substances help control body functions that involve the interaction of several organs. Insulin, a small protein, is an example of a hormone that regulates blood sugar. It involves the interaction of organs such as the pancreas and the liver. Secretin, is another example of a protein hormone. This substance assists in the digestive process by stimulating the pancreas and the intestine to create necessary digestive juices.

Enzymes

Enzymes are proteins that increase the rate of chemical reactions in the body. In fact, most of the necessary chemical reactions in the body would

Molecular Biology

not efficiently proceed without enzymes. For example, one type of enzyme functions as an aid in digesting large protein, carbohydrate and fat molecules into smaller molecules, while another assists the creation of DNA.

Transportation and Storage of Molecules

Protein is a major element in transportation of certain molecules. For example, hemoglobin is a protein that transports oxygen throughout the body. Protein is also sometimes used to store certain molecules. Ferritin is an example of a protein that combines with iron for storage in the liver.

Antibodies

Protein forms antibodies that help prevent infection, illness and disease. These proteins identify and assist in destroying antigens such as bacteria and viruses. They often work in conjunction with the other immune system cells. For example, these antibodies identify and then surround antigens in order to keep them contained until they can be destroyed by white blood cells.

Central Dogma of Molecular Biology

" The central dogma, enunciated by Crick in 1958 and the keystone of molecular biology ever since. is likely to prove will considerable . over-simplification."

THIS quotation is taken from the beginning of an unsigned articl headed" Central dogma reversed It. recountig the very important work of Dr Howard Temin' and others' showing that an RNA tumour virus can use viral RNA as a tem plate for DNA synthesis. This is not the first time that the idea of those central dogma has been mis-understood, in one way or another. In this article I explain why the term was originally introduced , its true meaning, and state why I think that, properly under-stood, it is still an idea of flmdamental im port ance. Thocentraldogma was put forward at a period when. much of what we now know in molecular genetics was not established. All we had to work on were certain fragmontary experimental results, themselves often rather uncertain and confused, and a bq,uudless optimism that the basic concepts involved were rather simple and vrobably much the same in all livinghings. In. such a situation well constructed theories can playa really useful partinstating problems clearly and thus guiding experiment.

Central Dogma : DNA to RNA to Protein

What is Central Dogma in Biology?

Central dogma explains how the DNA makes its own copies by DNA replication which then codes for the RNA by transcription and further, RNA codes for the proteins by translation.

Since the new DNA strands thus formed, have one strand of the parent DNA and the other is newly synthesized, the process is called semi-conservative DNA replication.

Transcription is the process by which the information is passed from one strand of the DNA to RNA. The DNA strand which undergoes the process consists of the parts, namely, promoter, structural gene and a terminator.

- The strand which synthesizes the RNA is called the template strand and the other strand is called the coding strand. The DNA-dependent RNA polymerase binds to promoter and catalyses the polymerization the 3' to 5' direction on the template strand.
- Once it reaches the terminator sequence, the process terminates and the newly synthesized RNA strand is released.
- The released RNA strand then undergoes post-transcriptional modifications.

Genetic code contains the information of the protein manufactured from RNA. We basically have four nitrogenous bases and three nucleotides together form a triplet codon which codes for one amino acid. Thus, the number possible amino acids range to 4 x 4 x 4 = 64 amino acids. But we have 20 naturally existing amino acids. This was explained by the features of genetic code, as per which some amino acids are coded by more than one codon thus making them degenerate. Each codon codes for only one specific amino acid and the codes are universal irrespective of the type of organism. Out of the 64 codons, 3 are stop codons which stop the process of transcription and one of the codon is an initiator codon i.e. AUG coding for Methionine.

Translation is the process by which the mRNA codes for a specific protein. It is an active process which requires the expense of ATP. This energy is provided by the charged tRNA. The complete machinery for translation is contained in the ribosomes. The ribosomes consist of a larger sub-unit and a smaller sub-unit. The larger sub-unit, in turn, consists of two tRNA molecules placed close enough so that peptide bond can be formed at the expense of enough energy. The mRNA enters the smaller sub-unit which is then held by the tRNA molecules of the complementary codon present in the larger sub-unit. Thus, two codons are held by two tRNA molecules placed close to each other and a peptide bond is formed between them. As this ocpress repeats, a long polypeptide chain of amino acids is synthesized.

Join BYJU's to learn the complete detailed procedure of the central dogma and get all your doubts cleared by our expert mentor's team.

Central dogma of molecular biology

The central dogma of molecular biology is an explanation of the flow of genetic information within a biological system. It was first stated by Francis Crick in 1958 and re-stated in a Nature paper published in 1970:[2]

The Central Dogma. This states that once 'information' has passed into protein it cannot get out again. In more detail, the transfer of information from nucleic acid to nucleic acid, or from nucleic acid to protein may be possible, but transfer from protein to protein, or from protein to nucleic acid is impossible. Information means here the precise determination of sequence, either of bases in the nucleic acid or of amino acid residues in the protein.

—?Francis Crick, 1958

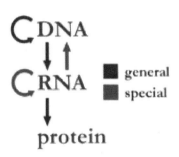

Information flow in biological systems

" The central dogma of molecular biology deals with the detailed residue-by-residue transfer of sequential information. It states that such information cannot be transferred back from protein to either protein or nucleic acid. "

—?Francis Crick

The central dogma has also been described as "DNA makes RNA and RNA makes protein,"[3] originally termed the sequence hypothesis and made as a positive statement by Crick. However, this simplification does not make it clear that the central dogma as stated by Crick does not preclude the reverse flow of information from RNA to DNA, only ruling out the flow from protein to RNA or DNA. Crick's use of the word dogma was unconventional, and has been controversial.

The dogma is a framework for understanding the transfer of sequence information between information-carrying biopolymers, in the most common or general case, in living organisms. There are 3 major classes of such biopolymers: DNA and RNA (both nucleic acids), and protein. There are 3×3 = 9 conceivable direct transfers of information that can occur between these. The dogma classes these into 3 groups of 3: three general transfers (believed to occur normally in most cells), three special transfers (known to occur, but only under specific conditions in case of some viruses or in a laboratory), and three unknown transfers (believed never to occur). The general transfers describe the normal flow of biological information: DNA can be copied to DNA (DNA replication), DNA information can be copied into mRNA (transcription), and proteins can be synthesized using the information in mRNA as a template (translation). The special transfers describe: RNA being copied from RNA (RNA replication), DNA being synthesised using an RNA template (reverse transcription), and proteins being synthesised directly from a DNA template without the use of mRNA. The unknown transfers describe: a protein being copied from a protein, synthesis of RNA using the primary structure of a protein as a template, and DNA synthesis using the primary structure of a protein as a template - these are not thought to naturally occur.

4

Concepts of Cell and Molecular Biology

INTRODUCTION

Trees in a forest, fish in a river, horseflies on a farm, lemurs in the jungle, reeds in a pond, worms in the soil — all these plants and animals are made of the building blocks we call cells. Like these examples, many living things consist of vast numbers of cells working in concert with one another. Other forms of life, however, are made of only a single cell, such as the many species of bacteria and protozoa. Cells, whether living on their own or as part of a multicellular organism, are usually too small to be seen without a light microscope.

Cells share many common features, yet they can look wildly different. In fact, cells have adapted over billions of years to a wide array of environments and functional roles. Nerve cells, for example, have long, thin extensions that can reach for meters and serve to transmit signals rapidly. Closely fitting, brick-shaped plant cells have a rigid outer layer that helps provide the structural support that trees and other plants require. Long, tapered muscle cells have an intrinsic stretchiness that allows them to change length within contracting and relaxing biceps.

Still, as different as these cells are, they all rely on the same basic strategies to keep the outside out, allow necessary substances in and permit others to leave, maintain their health, and replicate themselves. In fact, these traits are precisely what make a cell a cell.

Cells are the basic unit bounded by the membrane that consists of the fundamental molecules of life of which all living organisms are made up of. Cell Biology encompasses everything about cells from basic structure to functions of every cell organelle. A single cell is an organism itself such as yeast or bacterium other cells gain special functions soon after they mature.

These acquire unique functions as they mature. They cooperate with other cells and become the building blocks of multicellular organisms such as humans and animals. A group of small bacteria called mycoplasmas are the smallest known cells. Some of these unicellular organisms are spheres of about 0.3 micrometre in diameter with a mass of 10-14gram.

The cell (from Latin cella, meaning "small room" is the basic structural, functional, and biological unit of all known living organisms. A cell is the smallest unit of life that can replicate independently, and cells are often called the "building blocks of life". The study of cells is called cell biology.

Cells consist of cytoplasm enclosed within a membrane, which contains many biomolecules such as proteins and nucleic acids. Organisms can be classified as unicellular (consisting of a single cell; including bacteria) or multicellular (including plants and animals). While the number of cells in plants and animals varies from species to species, humans contain more than 10 trillion (1012) cells. Most plant and animal cells are visible only under a microscope, with dimensions between 1 and 100 micrometres.

The cell was discovered by Robert Hooke in 1665, who named the biological unit for its resemblance to cells inhabited by Christian monks in a monastery. Cell theory, first developed in 1839 by Matthias Jakob Schleiden and Theodor Schwann, states that all organisms are composed of one or more cells, that cells are the fundamental unit of structure and function in all living organisms, that all cells come from preexisting cells, and that all cells contain the hereditary information necessary for regulating cell functions and for transmitting information to the next generation of cells. Cells emerged on Earth at least 3.5 billion years ago.

Cellular and Molecular Biology

Cell and Molecular Biology is an interdisciplinary field of science that deals with the fields of chemistry, structure and biology as it seeks to understand life and cellular processes at the molecular level. Molecular cell Biology mainly focuses on the determination of cell fate and differentiation, growth regulation of cell, Cell adhesion and movement, Intracellular trafficking. The relationship of signalling to cellular growth and death, transcriptional regulation, mitosis, cellular differentiation and organogenesis, cell adhesion, motility and chemotaxis are more complex topics under Cellular and Molecular Biology. Molecular biology explores cells, their characteristics, parts, and chemical processes, and pays special attention to how molecules control a cell's activities and growth. The molecular components make up biochemical pathways that provide the cells with energy, facilitate processing "messages" from outside the cell itself, generate new proteins, and replicate the cellular DNA genome. To understand the behaviour of cells, it is important to add to the molecular level of description an understanding on the level of systems biology.

What is a cell?

The cell is the structural and functional unit of all living organisms, and is sometimes called the "building block of life." Some organisms, such as bacteria, are unicellular, consisting of a single cell. Other organisms, such as humans, are multicellular, (humans have an estimated 100 trillion cells; a typical cell size is 10 μm, a typical cell mass 1 nanogram). The largest known cell is an ostrich egg.

Each cell is at least somewhat self-contained and self-maintaining: it can take in nutrients, convert these nutrients into energy, carry out specialized functions, and reproduce as necessary. Each cell stores its own set of instructions for carrying out each of these activities. There are two types of cells, eukaryotic and prokaryotic.

Prokaryotic cells are usually singletons, while eukaryotic cells are usually found in multi-cellular organisms.

Cells are considered the basic units of life in part because they come in discrete and easily recognizable packages. That's because all cells are surrounded by a structure called the cell membrane — which, much like the walls of a house, serves as a clear boundary between the cell's internal and external environments. The cell membrane is sometimes also referred to as the plasma membrane.

Cell membranes are based on a framework of fat-based molecules called phospholipids, which physically prevent water-loving, or hydrophilic, substances from entering or escaping the cell. These membranes are also studded with proteins that serve various functions. Some of these proteins act as gatekeepers, determining what substances can and cannot cross the membrane. Others function as markers, identifying the cell as part of the same organism or as foreign. Still others work like fasteners, binding cells together so they can function as a unit. Yet other membrane proteins serve as communicators, sending and receiving signals from neighboring cells and the environment — whether friendly or alarming

Within this membrane, a cell's interior environment is water based. Called cytoplasm, this liquid environment is packed full of cellular machinery and structural elements. In fact, the concentrations of proteins inside a cell far outnumber those on the outside — whether the outside is ocean water (as in the case of a single-celled alga) or blood serum (as in the case of a red blood cell). Although cell membranes form natural barriers in watery environments, a cell must nonetheless expend quite a bit of energy to maintain the high concentrations of intracellular constituents necessary for its survival. Indeed, cells may use as much as 30 percent of their energy just to maintain the composition of their cytoplasm.

Cells are the basic building blocks of all living things. The human body is composed of trillions of cells. They provide structure for the body, take in nutrients from food, convert those nutrients into energy, and carry out specialized functions. Cells also contain the body's hereditary material and can make copies of themselves.

Cells have many parts, each with a different function. Some of these parts, called organelles, are specialized structures that perform certain tasks within the cell. Human cells contain the following major parts, listed in alphabetical order:

Cytoplasm

Within cells, the cytoplasm is made up of a jelly-like fluid (called the cytosol) and other structures that surround the nucleus.

Cytoskeleton

The cytoskeleton is a network of long fibers that make up the cell's structural framework. The cytoskeleton has several critical functions, including determining cell shape, participating in cell division, and allowing cells to move. It also provides a track-like system that directs the movement of organelles and other substances within cells.

Endoplasmic reticulum (ER)

This organelle helps process molecules created by the cell. The endoplasmic reticulum also transports these molecules to their specific destinations either inside or outside the cell.

Golgi apparatus

The Golgi apparatus packages molecules processed by the endoplasmic reticulum to be transported out of the cell.

Lysosomes and peroxisomes

These organelles are the recycling center of the cell. They digest foreign bacteria that invade the cell, rid the cell of toxic substances, and recycle worn-out cell components.

Mitochondria

Mitochondria are complex organelles that convert energy from food into a form that the cell can use. They have their own genetic material, separate from the DNA in the nucleus, and can make copies of themselves.

Concepts of Cell and Molecular Biology

Nucleus

The nucleus serves as the cell's command center, sending directions to the cell to grow, mature, divide, or die. It also houses DNA (deoxyribonucleic acid), the cell's hereditary material. The nucleus is surrounded by a membrane called the nuclear envelope, which protects the DNA and separates the nucleus from the rest of the cell.

Plasma membrane

The plasma membrane is the outer lining of the cell. It separates the cell from its environment and allows materials to enter and leave the cell.

Ribosomes

Ribosomes are organelles that process the cell's genetic instructions to create proteins. These organelles can float freely in the cytoplasm or be connected to the endoplasmic reticulum.

Basic Concepts of Cell Biology

The main aim of this course is for students to understand the organisation and functioning of the cell, including the principles of different signal transduction mechanisms. By the end of this course, students will able to decribe relations between DNA, RNA and protein synthesis, between structure, specific location and function of proteins or organelles, and between the sensation of stimuli and control of cellular metabolic reactions, growth and mitosis/meiosis. Along the course, students acquire the scientifically correct terminology to denote structural features of macromolecules, modes of signalling mechanisms and modulles as well as regulatory cellular processes.

Other Components of Cells

As previously mentioned, a cell's cytoplasm is home to numerous functional and structural elements. These elements exist in the form of molecules and organelles — picture them as the tools, appliances, and inner rooms of the cell. Major classes of intracellular organic molecules include nucleic acids, proteins, carbohydrates, and lipids, all of which are essential to the cell's functions.

Nucleic acids are the molecules that contain and help express a cell's genetic code. There are two major classes of nucleic acids: deoxyribonucleic acid (DNA) and ribonucleic acid (RNA). DNA is the molecule that contains all of the information required to build and maintain the cell; RNA has several roles associated with expression of the information stored in DNA. Of course, nucleic acids alone aren't responsible for the preservation and expression of genetic material: Cells also use proteins to help replicate the genome and accomplish the profound structural changes that underlie cell division.

Proteins are a second type of intracellular organic molecule. These substances are made from chains of smaller molecules called amino acids, and they serve a variety of functions in the cell, both catalytic and structural. For example, proteins called enzymes convert cellular molecules (whether proteins, carbohydrates, lipids, or nucleic acids) into other forms that might help a cell meet its energy needs, build support structures, or pump out wastes.

Carbohydrates, the starches and sugars in cells, are another important type of organic molecule. Simple carbohydrates are used for the cell's immediate energy demands, whereas complex carbohydrates serve as intracellular energy stores. Complex carbohydrates are also found on a cell's surface, where they play a crucial role in cell recognition.

Finally, lipids or fat molecules are components of cell membranes — both the plasma membrane and various intracellular membranes. They are also involved in energy storage, as well as relaying signals within cells and from the bloodstream to a cell's interior.

Some cells also feature orderly arrangements of molecules called organelles. Similar to the rooms in a house, these structures are partitioned off from the rest of a cell's interior by their own intracellular membrane. Organelles contain highly technical equipment required for specific jobs within the cell. One example is the mitochondrion — commonly known as the cell's "power plant" — which is the organelle that holds and maintains the machinery involved in energy-producing chemical reactions.

What Are the Different Categories of Cells?

Rather than grouping cells by their size or shape, scientists typically categorize them by how their genetic material is packaged. If the DNA within a cell is not separated from the cytoplasm, then that cell is a prokaryote. All known prokaryotes, such as bacteria and archaea, are single cells. In contrast, if the DNA is partitioned off in its own membrane-bound room called the nucleus, then that cell is a eukaryote. Some eukaryotes, like amoebae, are free-living, single-celled entities. Other eukaryotic cells are part of multicellular organisms. For instance, all plants and animals are made of eukaryotic cells — sometimes even trillions of them.

The Function & Nature of Cells

- A cell is bounded by a plasma membrane that forms a selective barrier allows nutrients to enter and leave waste products.

- The cell interior is organised into different special compartments or organelles surrounded by a separate membrane.
- The nucleus(major organelle) holds genetic information necessary for reproduction and cell growth.
- Every cell has one nucleus and other types of organelles exist in multiple copies in the cytoplasm.
- Organelles involve mitochondria responsible for the energy transactions vital for cell survival.
- Lysosomes digest unwanted materials in the cell.
- Endoplasmic reticulum plays a significant role in the internal organisation of the cell by synthesising selective molecules and processing, directing and sorting them to their appropriate locations.

Cell Biology

Cell biology is the study of cell structure and function, and it revolves around the concept that the cell is the fundamental unit of life. Focusing on the cell permits a detailed understanding of the tissues and organisms that cells compose. Some organisms have only one cell, while others are organized into cooperative groups with huge numbers of cells. On the whole, cell biology focuses on the structure and function of a cell, from the most general properties shared by all cells, to the unique, highly intricate functions particular to specialized cells.

The starting point for this discipline might be considered the 1830s. Though scientists had been using microscopes for centuries, they were not always sure what they were looking at. Robert Hooke's initial observation in 1665 of plant-cell walls in slices of cork was followed shortly by Antonie van Leeuwenhoek's first descriptions of live cells with visibly moving parts. In the 1830s two scientists who were colleagues — Schleiden, looking at plant cells, and Schwann, looking first at animal cells — provided the first clearly stated definition of the cell. Their definition stated that that all living creatures, both simple and complex, are made out of one or more cells, and the cell is the structural and functional unit of life — a concept that became known as cell theory.

As microscopes and staining techniques improved over the nineteenth and twentieth centuries, scientists were able to see more and more internal detail within cells. The microscopes used by van Leeuwenhoek probably magnified specimens a few hundredfold. Today high-powered electron microscopes can magnify specimens more than a million times and can reveal the shapes of organelles at the scale of a micrometer and below. With confocal microscopy a series of images can be combined, allowing researchers to generate detailed three-dimensional representations of cells. These improved imaging techniques have helped us better understand the wonderful complexity of cells and the structures they form.

There are several main subfields within cell biology. One is the study of cell energy and the biochemical mechanisms that support cell metabolism. As cells are machines unto themselves, the focus on cell energy overlaps with the pursuit of questions of how energy first arose in original primordial cells, billions of years ago. Another subfield of cell biology concerns the genetics of the cell and its tight interconnection with the proteins controlling the release of genetic information from the nucleus to the cell cytoplasm. Yet another subfield focuses on the structure of cell components, known as subcellular compartments. Cutting across many biological disciplines is the additional subfield of cell biology, concerned with cell communication and signaling, concentrating on the messages that cells give to and receive from other cells and themselves. And finally, there is the subfield primarily concerned with the cell cycle, the rotation of phases beginning and ending with cell division and focused on different periods of growth and DNA replication. Many cell biologists dwell at the intersection of two or more of these subfields as our ability to analyze cells in more complex ways expands.

In line with continually increasing interdisciplinary study, the recent emergence of systems biology has affected many biological disciplines; it is a methodology that encourages the analysis of living systems within the context of other systems. In the field of cell biology, systems biology has enabled the asking and answering of more complex questions, such as the interrelationships of gene regulatory networks, evolutionary relationships between genomes, and the interactions between intracellular signaling networks. Ultimately, the broader a lens we take on our discoveries in cell biology, the more likely we can decipher the complexities of all living systems, large and small.

WHAT IS CELL BIOLOGY?

Cell biology is the subdiscipline of biology that studies the basic unit of life, the cell. It deals with all aspects of the cell including cell anatomy, cell division (mitosis and meiosis), and cell processes including cell respiration, and cell death. Cell biology does not stand alone as a discipline but is closely related to other areas of biology such as genetics, molecular biology, and biochemistry.

Based on one of the basic principles of biology, the cell theory, the study of cells would not have been possible without the invention of the microscope. With the advanced microscopes of today, such as the Scanning Electron

Concepts of Cell and Molecular Biology

Microscope and Transmission Electron Microscope, cell biologists are able to obtain detailed images of the smallest of cell structures and organelles.

The cell is the basic structural, functional, and biological unit of all known living organisms. A cell is the smallest unit of life that can replicate independently, and cells are often called the "building blocks of life". The study of cells is called cell biology.

Cells consist of cytoplasm enclosed within a membrane, which contains many biomolecules such as proteins and nucleic acids.[2] Organisms can be classified as unicellular (consisting of a single cell; including bacteria) or multicellular (including plants and animals). While the number of cells in plants and animals varies from species to species, humans contain more than 10 trillion (10^{12}) cells.[3] Most plant and animal cells are visible only under a microscope, with dimensions between 1 and 100 micrometres.[4]

The cell was discovered by Robert Hooke in 1665, who named the biological unit for its resemblance to cells inhabited by Christian monks in a monastery.[5][6] Cell theory, first developed in 1839 by Matthias Jakob Schleiden and Theodor Schwann, states that all organisms are composed of one or more cells, that cells are the fundamental unit of structure and function in all living organisms, that all cells come from preexisting cells, and that all cells contain the hereditary information necessary for regulating cell functions and for transmitting information to the next generation of cells.[7]Cells emerged on Earth at least 3.5 billion years ago.[

What is Molecular Biology?

Molecular biology is a branch of science concerning biological activity at the molecular level.

The field of molecular biology overlaps with biology and chemistry and in particular, genetics and biochemistry. A key area of molecular biology concerns understanding how various cellular systems interact in terms of the way DNA, RNA and protein synthesis function.

The specific techniques used in molecular biology are native to the field but may also be combined with methods and concepts concerning genetics and biochemistry, so there is no big distinction made between these disciplines.

However, when the fields are considered independently of each other, biochemistry concerns chemical materials and essential processes that take place in living organisms. The role, function and structure of biomolecules are key areas of focus among biochemists, as is the chemistry behind biological functions and the production of biomolecules.

Genetics is concerned with the effects of genes on living organisms, which are often examined through "knock-out" studies, where animal models are designed so that they lack one or more genes compared to a "wild type" or regular phenotype.

Molecular biology looks at the molecular mechanisms behind processes such as replication, transcription, translation and cell function. One way to describe the basis of molecular biology is to say it concerns understanding how genes are transcribed into RNA and how RNA is then translated into protein. However, this simplified picture is currently be reconsidered and revised due to new discoveries concerning the roles of RNA.

Much like mini-fiefdoms, cells have all the equipment and expertise necessary to carry out the functions of life. A cell can eat, grow, and move. It can perform necessary maintenance, recycle parts, and dispose of wastes. It can adapt to changes in its environment; and it can even replicate itself.

Despite these similarities, all cells are not equal. Some are truly self-sustaining, as with single-celled bacteria or yeast, whereas others live communally, sometimes as part of complex multicellular organisms. Cells also differ in size. Although cells can be quite large — consider a frog's egg, for example — most are too small to see with the naked eye. Indeed, the development of light microscopy was essential to man's discovery of cells.

Don't be lulled by familiar schematic drawings of oval-shaped cells, either. Real cells are three-dimensional, of course, and they exist in a variety of intricate and remarkable shapes. For instance, a single human nerve cell can be over one meter long, extending from your backbone to your big toe. Compare that with the cells that line your small intestine, which have dozens of tiny, fingerlike projections to maximize the surface area across which nutrients can pass.

What Are the Essential Characteristics of Cells?

Cells Are the Basic Units of Living Organisms

Trees in a forest, fish in a river, horseflies on a farm, lemurs in the jungle, reeds in a pond, worms in the soil — all these plants and animals are made of the building blocks we call cells. Like these examples, many living things consist of vast numbers of cells working in concert with one another. Other forms of life, however, are made of only a single cell, such as the many species of bacteria and protozoa. Cells, whether living on their own or as part of a multicellular organism, are usually too small to be seen without a light microscope.

Cells share many common features, yet they can look wildly different. In fact, cells have adapted over billions of years to a wide array of environments and functional roles. Nerve cells, for example, have long, thin extensions that can reach for meters and serve to transmit signals rapidly. Closely fitting,

Concepts of Cell and Molecular Biology

brick-shaped plant cells have a rigid outer layer that helps provide the structural support that trees and other plants require. Long, tapered muscle cells have an intrinsic stretchiness that allows them to change length within contracting and relaxing biceps.

Still, as different as these cells are, they all rely on the same basic strategies to keep the outside out, allow necessary substances in and permit others to leave, maintain their health, and replicate themselves. In fact, these traits are precisely what make a cell a cell.

How Did Cells Originate?

Researchers hypothesize that all organisms on Earth today originated from a single cell that existed some 3.5 to 3.8 billion years ago. This original cell was likely little more than a sac of small organic molecules and RNA-like material that had both informational and catalytic functions. Over time, the more stable DNA molecule evolved to take over the information storage function, whereas proteins, with a greater variety of structures than nucleic acids, took over the catalytic functions.

As described in the previous section, the absence or presence of a nucleus — and indeed, of all membrane-bound organelles — is important enough to be a defining feature by which cells are categorized as either prokaryotes or eukaryotes. Scientists believe that the appearance of self-contained nuclei and other organelles represents a major advance in the evolution of cells. But where did these structures come from? More than one billion years ago, some cells "ate" by engulfing objects that floated in the liquid environment in which they existed. Then, according to some theories of cellular evolution, one of the early eukaryotic cells engulfed a prokaryote, and together the two cells formed a symbiotic relationship. In particular, the engulfed cell began to function as an organelle within the larger eukaryotic cell that consumed it. Both chloroplasts and mitochondria, which exist in modern eukaryotic cells and still retain their own genomes, are thought to have arisen in this manner

Photosynthetic Cells Capture Light Energy and Convert It to Chemical Energy

Cells get nutrients from their environment, but where do those nutrients come from? Virtually all organic material on Earth has been produced by cells that convert energy from the Sun into energy-containing macromolecules. This process, called photosynthesis, is essential to the global carbon cycle and organisms that conduct photosynthesis represent the lowest level in most food chains

What Is Photosynthesis? Why Is it Important?

Most living things depend on photosynthetic cells to manufacture the complex organic molecules they require as a source of energy. Photosynthetic cells are quite diverse and include cells found in green plants, phytoplankton, and cyanobacteria. During the process of photosynthesis, cells use carbon dioxide and energy from the Sun to make sugar molecules and oxygen. These sugar molecules are the basis for more complex molecules made by the photosynthetic cell, such as glucose. Then, via respiration processes, cells use oxygen and glucose to synthesize energy-rich carrier molecules, such as ATP, and carbon dioxide is produced as a waste product. Therefore, the synthesis of glucose and its breakdown by cells are opposing processes.

The building and breaking of carbon-based material — from carbon dioxide to complex organic molecules (photosynthesis) then back to carbon dioxide (respiration) — is part of what is commonly called the global carbon cycle. Indeed, the fossil fuels we use to power our world today are the ancient remains of once-living organisms, and they provide a dramatic example of this cycle at work. The carbon cycle would not be possible without photosynthesis, because this process accounts for the "building" portion of the cycle (Figure 2).

However, photosynthesis doesn't just drive the carbon cycle — it also creates the oxygen necessary for respiring organisms. Interestingly, although green plants contribute much of the oxygen in the air we breathe, phytoplankton and cyanobacteria in the world's oceans are thought to produce between one-third and one-half of atmospheric oxygen on Earth.

What Cells and Organelles Are Involved in Photosynthesis?

Photosynthetic cells contain special pigments that absorb light energy. Different pigments respond to different wavelengths of visible light. Chlorophyll, the primary pigment used in photosynthesis, reflects green light and absorbs red and blue light most strongly. In plants, photosynthesis takes place in chloroplasts, which contain the chlorophyll. Chloroplasts are surrounded by a double membrane and contain a third inner membrane, called the thylakoid membrane, that forms long folds within the organelle. In electron micrographs, thylakoid membranes look like stacks of coins, although the compartments they form are connected like a maze of chambers. The green pigment chlorophyll is located within the thylakoid membrane, and the space between the thylakoid and the chloroplast membranes is called the stroma (Figure 3, Figure 4).

Chlorophyll A is the major pigment used in photosynthesis, but there are several types of chlorophyll and numerous other pigments that respond to light, including red, brown, and blue pigments. These other pigments may help channel light energy to chlorophyll A or protect the cell from photodamage. For example, the photosynthetic protists called dinoflagellates, which are responsible for the "red tides" that often prompt warnings against eating shellfish, contain a variety of light-sensitive pigments, including both chlorophyll and the red pigments responsible for their dramatic coloration.

What Are the Steps of Photosynthesis?

Photosynthesis consists of both light-dependent reactions and light-independent reactions. In plants, the so-called "light" reactions occur within the chloroplast thylakoids, where the aforementioned chlorophyll pigments reside. When light energy reaches the pigment molecules, it energizes the electrons within them, and these electrons are shunted to an electron transport chain in the thylakoid membrane. Every step in the electron transport chain then brings each electron to a lower energy state and harnesses its energy by producing ATP and NADPH. Meanwhile, each chlorophyll molecule replaces its lost electron with an electron from water; this process essentially splits water molecules to produce oxygen.

Once the light reactions have occurred, the light-independent or "dark" reactions take place in the chloroplast stroma. During this process, also known as carbon fixation, energy from the ATP and NADPH molecules generated by the light reactions drives a chemical pathway that uses the carbon in carbon dioxide (from the atmosphere) to build a three-carbon sugar called glyceraldehyde-3-phosphate (G3P). Cells then use G3P to build a wide variety of other sugars (such as glucose) and organic molecules. Many of these interconversions occur outside the chloroplast, following the transport of G3P from the stroma. The products of these reactions are then transported to other parts of the cell, including the mitochondria, where they are broken down to make more energy carrier molecules to satisfy the metabolic demands of the cell. In plants, some sugar molecules are stored as sucrose or starch.

CONCLUSION

Photosynthetic cells contain chlorophyll and other light-sensitive pigments that capture solar energy. In the presence of carbon dioxide, such cells are able to convert this solar energy into energy-rich organic molecules, such as glucose. These cells not only drive the global carbon cycle, but they also produce much of the oxygen present in atmosphere of the Earth. Essentially, nonphotosynthetic cells use the products of photosynthesis to do the opposite of photosynthesis: break down glucose and release carbon dioxide.

Metabolism is the Complete Set of Biochemical Reactions within a Cell

A cell's daily operations are accomplished through the biochemical reactions that take place within the cell. Reactions are turned on and off or sped up and slowed down according to the cell's immediate needs and overall functions. At any given time, the numerous pathways involved in building up and breaking down cellular components must be monitored and balanced in a coordinated fashion. To achieve this goal, cells organize reactions into various enzyme-powered pathways.

What Do Enzymes Do?

Enzymes are protein catalysts that speed biochemical reactions by facilitating the molecular rearrangements that support cell function. Recall that chemical reactions convert substrates into products, often by attaching chemical groups to or breaking off chemical groups from the substrates. For example, in the final step of glycolysis, an enzyme called pyruvate kinase transfers a phosphate group from one substrate (phosphoenolpyruvate) to another substrate (ADP), thereby generating pyruvate and ATP as products (Figure 1).

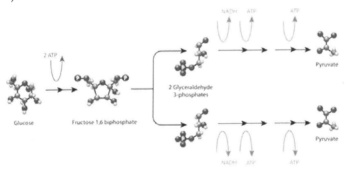

Figure 1: Glycolysis

Energy is used to convert glucose to a 6 carbon form. Thereafter, energy is generated to create two molecules of pyruvate.

Enzymes are flexible proteins that change shape when they bind with substrate molecules. In fact, this binding and shape changing ability is how enzymes manage to increase reaction rates. In many cases, enzymes function by bringing two substrates into close proximity and orienting them for easier electron transfer. Shape or conformational changes can also act as an on/off switch. For example, when inhibitor molecules bind to a site on an enzyme distinct from the substrate site, they can make the enzyme assume an inactive conformation, thereby preventing it from catalyzing a reaction.

Conversely, the binding of activator molecules can make an enzyme assume an active conformation, essentially turning it on (Figure 2).

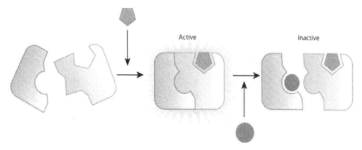

Figure 2: Activation and inactivation of of enzyme reaction

Enzymes are proteins that can change shape and therefore become active or inactive. An activator molecule (green pentagon) can bind to an enzyme (light green puzzle shape) and change its overall shape. Note the transformation of the triangular point on the green enzyme into a rounded shape. This transformation enables the enzyme to better bind with its substrate (light pink puzzle piece). In contrast, an inhibitor molecule (pink circle) can prevent the interaction of an enzyme with its substrate and render it inactive.

What Are Metabolic Pathways?

Many of the molecular transformations that occur within cells require multiple steps to accomplish. Recall, for instance, that cells split one glucose molecule into two pyruvate molecules by way of a ten-step process called glycolysis. This coordinated series of chemical reactions is an example of a metabolic pathway in which the product of one reaction becomes the substrate for the next reaction. Consequently, the intermediate products of a metabolic pathway may be short-lived.

Sometimes, the enzymes involved in a particular metabolic pathway are physically connected, allowing the products of one reaction to be efficiently channeled to the next enzyme in the pathway. For example, pyruvate dehydrogenase is a complex of three different enzymes that catalyze the path from pyruvate (the end product of glycolysis) to acetyl CoA (the first substrate in the citric acid cycle). Within this complex, intermediate products are passed directly from one enzyme to the next.

How Do Cells Keep Chemical Reactions in Balance?

Cells are expert recyclers. They disassemble large molecules into simpler building blocks and then use those building blocks to create the new components they require. The breaking down of complex organic molecules occurs via catabolic pathways and usually involves the release of energy. Through catabolic pathways, polymers such as proteins, nucleic acids, and polysaccharides are reduced to their constituent parts: amino acids, nucleotides, and sugars, respectively. In contrast, the synthesis of new macromolecules occurs via anabolic pathways that require energy input (Figure 4).

Cells must balance their catabolic and anabolic pathways in order to control their levels of critical metabolites — those molecules created by enzymatic activity — and ensure that sufficient energy is available. For example, if supplies of glucose start to wane, as might happen in the case of starvation, cells will synthesize glucose from other materials or start sending fatty acids into the citric acid cycle to generate ATP. Conversely, in times of plenty, excess glucose is converted into storage forms, such as glycogen, starches, and fats.

How Do Cells Manage All Their Chemical Reactions?

Not only do cells need to balance catabolic and anabolic pathways, but they must also monitor the needs and surpluses of all their different metabolic pathways. In order to bolster a particular pathway, cells can increase the amount of a necessary (rate-limiting) enzyme or use activators to convert that enzyme into an active conformation. Conversely, to slow down or halt a pathway, cells can decrease the amount of an enzyme or use inhibitors to make the enzyme inactive.

Such up- and down-regulation of metabolic pathways is often a response to changes in concentrations of key metabolites in the cell. For example, a cell may take stock of its levels of intermediate metabolites and tune the glycolytic pathway and the synthesis of glucose accordingly. In some instances, the products of a metabolic pathway actually serve as inhibitors of their own synthesis, in a process known as feedback inhibition (Figure 5). For example, the first intermediate in glycolysis, glucose-6-phosphate, inhibits the very enzyme that produces it, hexokinase.

Conclusion

The management of biochemical reactions with enzymes is an important part of cellular maintenance. Enzymatic activity allows a cell to respond to changing environmental demands and regulate its metabolic pathways, both of which are essential to cell survival.

How Do Cells Decode Genetic Information into Functional Proteins?

Cells archive this information in DNA, which serves as a master set of instructions for building proteins. It is a beautiful system made complex by many levels of control, on-off switches, feedback, and fine-tuning. Segments of DNA are transcribed into RNA, and this RNA is then translated into proteins. The resulting proteins then fold into their three-dimensional configurations and combine with other proteins, or are decorated with sugars or fats to create finely-crafted tools for carrying out specific cellular functions. Protein functions range from structural supports and motors to catalysts of biochemical reactions and monitors of the cell's internal and external environments.

Every step in the protein production pathway can be adjusted up or down as the cell's needs dictate. The ability to carefully regulate transcription, translation, protein folding and modification, and protein function is a feature that makes cells such resilient and versatile life-forms.

Information Transfer in Cells Requires Many Proteins and Nucleic Acids

The genetic information stored in DNA is a living archive of instructions that cells use to accomplish the functions of life. Inside each cell, catalysts seek out the appropriate information from this archive and use it to build new proteins — proteins that make up the structures of the cell, run the biochemical reactions in the cell, and are sometimes manufactured for export. Although all of the cells that make up a multicellular organism contain identical genetic information, functionally different cells within the organism use different sets of catalysts to express only specific portions of these instructions to accomplish the functions of life.

How Is Genetic Information Passed on in Dividing Cells?

When a cell divides, it creates one copy of its genetic information — in the form of DNA molecules — for each of the two resulting daughter cells. The accuracy of these copies determines the health and inherited features of the nascent cells, so it is essential that the process of DNA replication be as accurate as possible.

One factor that helps ensure precise replication is the double-helical structure of DNA itself. In particular, the two strands of the DNA double helix are made up of combinations of molecules called nucleotides. DNA is constructed from just four different nucleotides — adenine (A), thymine (T), cytosine (C), and guanine (G) — each of which is named for the nitrogenous base it contains. Moreover, the nucleotides that form one strand of the DNA double helix always bond with the nucleotides in the other strand according to a pattern known as complementary base-pairing — specifically, A always pairs with T, and C always pairs with G (Figure 2). Thus, during cell division, the paired strands unravel and each strand serves as the template for synthesis of a new complementary strand.

In most multicellular organisms, every cell carries the same DNA, but this genetic information is used in varying ways by different types of cells. In other words, what a cell "does" within an organism dictates which of its genes are expressed. Nerve cells, for example, synthesize an abundance of chemicals called neurotransmitters, which they use to send messages to other cells, whereas muscle cells load themselves with the protein-based filaments necessary for muscle contractions.

What Are the Initial Steps in Accessing Genetic Information?

Transcription is the first step in decoding a cell's genetic information. During transcription, enzymes called RNA polymerases build RNA molecules that are complementary to a portion of one strand of the DNA double helix (Figure 3).

RNA molecules differ from DNA molecules in several important ways: They are single stranded rather than double stranded; their sugar component is a ribose rather than a deoxyribose; and they include uracil (U) nucleotides rather than thymine (T) nucleotides (Figure 4). Also, because they are single strands, RNA molecules don't form helices; rather, they fold into complex structures that are stabilized by internal complementary base-pairing.

Three general classes of RNA molecules are involved in expressing the genes encoded within a cell's DNA. Messenger RNA (mRNA) molecules carry the coding sequences for protein synthesis and are called transcripts; ribosomal RNA (rRNA) molecules form the core of a cell's ribosomes (the structures in which protein synthesis takes place); and transfer RNA (tRNA) molecules carry amino acids to the ribosomes during protein synthesis. In eukaryotic cells, each class of RNA has its own polymerase, whereas in prokaryotic cells, a single RNA polymerase synthesizes the different class of RNA. Other types of RNA also exist but are not as well understood, although they appear to play regulatory roles in gene expression and also be involved in protection against invading viruses.

mRNA is the most variable class of RNA, and there are literally thousands of different mRNA molecules present in a cell at any given time. Some mRNA molecules are abundant, numbering in the hundreds or thousands, as is often true of transcripts encoding structural proteins. Other mRNAs are quite rare, with perhaps only a single copy present, as is sometimes the case for transcripts that encode signaling proteins. mRNAs also vary in how long-lived they are. In eukaryotes, transcripts for structural proteins may remain intact for over ten hours, whereas transcripts for signaling proteins may be degraded in less than ten minutes.

Cells can be characterized by the spectrum of mRNA molecules present within them; this spectrum is called the transcriptome. Whereas each cell in

Concepts of Cell and Molecular Biology 103

a multicellular organism carries the same DNA or genome, its transcriptome varies widely according to cell type and function. For instance, the insulin-producing cells of the pancreas contain transcripts for insulin, but bone cells do not. Even though bone cells carry the gene for insulin, this gene is not transcribed. Therefore, the transcriptome functions as a kind of catalog of all of the genes that are being expressed in a cell at a particular point in time.

What Is the Function of Ribosomes?

Ribosomes are the sites in a cell in which protein synthesis takes place. Cells have many ribosomes, and the exact number depends on how active a particular cell is in synthesizing proteins. For example, rapidly growing cells usually have a large number of ribosomes (Figure 5).

Ribosomes are complexes of rRNA molecules and proteins, and they can be observed in electron micrographs of cells. Sometimes, ribosomes are visible as clusters, called polyribosomes. In eukaryotes (but not in prokaryotes), some of the ribosomes are attached to internal membranes, where they synthesize the proteins that will later reside in those membranes, or are destined for secretion (Figure 6). Although only a few rRNA molecules are present in each ribosome, these molecules make up about half of the ribosomal mass. The remaining mass consists of a number of proteins — nearly 60 in prokaryotic cells and over 80 in eukaryotic cells.

Within the ribosome, the rRNA molecules direct the catalytic steps of protein synthesis — the stitching together of amino acids to make a protein molecule. In fact, rRNA is sometimes called a ribozyme or catalytic RNA to reflect this function.

Eukaryotic and prokaryotic ribosomes are different from each other as a result of divergent evolution. These differences are exploited by antibiotics, which are designed to inhibit the prokaryotic ribosomes of infectious bacteria without affecting eukaryotic ribosomes, thereby not interfering with the cells of the sick host.

How Does the Whole Process Result in New Proteins?

After the transcription of DNA to mRNA is complete, translation — or the reading of these mRNAs to make proteins — begins. Recall that mRNA molecules are single stranded, and the order of their bases — A, U, C, and G — is complementary to that in specific portions of the cell's DNA. Each mRNA dictates the order in which amino acids should be added to a growing protein as it is synthesized. In fact, every amino acid is represented by a three-nucleotide sequence or codon along the mRNA molecule. For example, AGC is the mRNA codon for the amino acid serine, and UAA is a signal to stop translating a protein — also called the stop codon .

Molecules of tRNA are responsible for matching amino acids with the appropriate codons in mRNA. Each tRNA molecule has two distinct ends, one of which binds to a specific amino acid, and the other which binds to the corresponding mRNA codon. During translation, these tRNAs carry amino acids to the ribosome and join with their complementary codons. Then, the assembled amino acids are joined together as the ribosome, with its resident rRNAs, moves along the mRNA molecule in a ratchet-like motion. The resulting protein chains can be hundreds of amino acids in length, and synthesizing these molecules requires a huge amount of chemical energy (Figure 8).

In prokaryotic cells, transcription (DNA to mRNA) and translation (mRNA to protein) are so closely linked that translation usually begins before transcription is complete. In eukaryotic cells, however, the two processes are separated in both space and time: mRNAs are synthesized in the nucleus, and proteins are later made in the cytoplasm.

Conclusion

Cellular DNA contains instructions for building the various proteins the cell needs to survive. In order for a cell to manufacture these proteins, specific genes within its DNA must first be transcribed into molecules of mRNA; then, these transcripts must be translated into chains of amino acids, which later fold into fully functional proteins. Although all of the cells in a multicellular organism contain the same set of genetic information, the transcriptomes of different cells vary depending on the cells' structure and function in the organism.

Proteins Are Responsible for a Diverse Range of Structural and Catalytic Functions in Cells

The collection of proteins within a cell determines its health and function. Proteins are responsible for nearly every task of cellular life, including cell shape and inner organization, product manufacture and waste cleanup, and routine maintenance. Proteins also receive signals from outside the cell and mobilize intracellular response. They are the workhorse macromolecules of the cell and are as diverse as the functions they serve.

How Diverse Are Proteins?

Proteins can be big or small, mostly hydrophilic or mostly hydrophobic, exist alone or as part of a multi-unit structure, and change shape frequently or remain virtually immobile. All of these differences arise from the unique amino acid sequences that make up proteins. Fully folded proteins also have distinct surface characteristics that determine which other molecules they interact with. When proteins bind with other molecules, their conformation can change in subtle or dramatic ways.

Not surprisingly, protein functions are as diverse as protein structures. For example, structural proteins maintain cell shape, akin to a skeleton, and they compose structural elements in connective tissues like cartilage and bone in vertebrates. Enzymes are another type of protein, and these molecules catalyze the biochemical reactions that occur in cells. Yet other proteins work as monitors, changing their shape and activity in response to metabolic signals or messages from outside the cell. Cells also secrete various proteins that become part of the extracellular matrix or are involved in intercellular communication.

Proteins are sometimes altered after translation and folding are complete. In such cases, so-called transferase enzymes add small modifier groups, such as phosphates or carboxyl groups, to the protein. These modifications often shift protein conformation and act as molecular switches that turn the activity of a protein on or off. Many post-translational modifications are reversible, although different enzymes catalyze the reverse reactions. For example, enzymes called kinases add phosphate groups to proteins, but enzymes called phosphatases are required to remove these phosphate groups (Figure 1).

How Do Proteins Provide Structural Support for Cells?

The cytoplasm is highly structured, thanks to proteins. Particularly in eukaryotic cells, which tend to be bigger and need more mechanical support than prokaryotic cells, an extensive network of filaments — microtubules, actin filaments, and intermediate filaments — can be detected with a variety of microscopic methods. Microtubules play a major role in organizing the cytoplasm and in the distribution of organelles. They also form the mitotic spindle during cell division. Actin filaments are involved in various forms of cell movement, including cell locomotion, contraction of muscle cells, and cell division (Figure 2). Intermediate filaments are strong fibers that serve as architectural supports inside cells.

How Do Proteins Aid the Biochemical Reactions of a Cell?

Cells rely on thousands of different enzymes to catalyze metabolic reactions. Enzymes are proteins, and they make a biochemical reaction more likely to proceed by lowering the activation energy of the reaction, thereby making these reactions proceed thousands or even millions of times faster than they would without a catalyst. Enzymes are highly specific to their substrates. They bind these substrates at complementary areas on their surfaces, providing a snug fit that many scientists compare to a lock and key. Enzymes work by binding one or more substrates, bringing them together so that a reaction can take place, and releasing them once the reaction is complete. In particular, when substrate binding occurs, enzymes undergo a conformational shift that orients or strains the substrates so that they are more reactive (Figure 3).

The name of an enzyme usually refers to the type of biochemical reaction it catalyzes. For example, proteases break down proteins, and dehydrogenases oxidize a substrate by removing hydrogen atoms. As a general rule, the "-ase" suffix identifies a protein as an enzyme, whereas the first part of an enzyme's name refers to the reaction that it catalyzes.

What Do Proteins Do in the Plasma Membrane?

The proteins in the plasma membrane typically help the cell interact with its environment. For example, plasma membrane proteins carry out functions as diverse as ferrying nutrients across the plasma membrane, receiving chemical signals from outside the cell, translating chemical signals into intracellular action, and sometimes anchoring the cell in a particular location (Figure 4).

The overall surfaces of membrane proteins are mosaics, with patches of hydrophobic amino acids where the proteins contact lipids in the membrane bilayer and patches of hydrophilic amino acids on the surfaces that extend into the water-based cytoplasm. Many proteins can move within the plasma membrane through a process called membrane diffusion. This concept of membrane-bound proteins that can travel within the membrane is called the fluid-mosaic model of the cell membrane. The portions of membrane proteins that extend beyond the lipid bilayer into the extracellular environment are also hydrophilic and are frequently modified by the addition of sugar molecules. Other proteins are associated with the membrane but not inserted into it. They are sometimes anchored to lipids in the membrane or bound to other membrane proteins (Figure 5).

Conclusion

Proteins serve a variety of functions within cells. Some are involved in structural support and movement, others in enzymatic activity, and still others in interaction with the outside world. Indeed, the functions of individual proteins are as varied as their unique amino acid sequences and complex three-dimensional physical structures.

How Are Eukaryotic Cells Organized into Smaller Parts?

Scientists aren't just interested in the individual molecules found in cells or in cells' metabolic functions and pathways; they also seek to learn more about the ways in which larger groupings of molecules serve cells. These groupings take on many forms and are responsible for a wide variety of functions. For instance, a cell's membranes make up its outer boundary and partition off its organelles, and its cytoskeleton provides three-dimensional support and the means for movement. Other specialized structures, such as mitochondria, chloroplasts, and cell walls, have evolved to carry out vital functions, but they are found only in specific categories of cells. Therefore, understanding these structures not only provides insight into cellular function, but it also helps elucidate the differences between various types of organisms.

Specialized Membranes Organize the Eukaryotic Cell Cytoplasm into Compartments

Cell membranes protect and organize cells. All cells have an outer plasma membrane that regulates not only what enters the cell, but also how much of any given substance comes in. Unlike prokaryotes, eukaryotic cells also possess internal membranes that encase their organelles and control the exchange of essential cell components. Both types of membranes have a specialized structure that facilitates their gatekeeping function.

What Are Cellular Membranes Made Of?

With few exceptions, cellular membranes — including plasma membranes and internal membranes — are made of glycerophospholipids, molecules composed of glycerol, a phosphate group, and two fatty acid chains. Glycerol is a three-carbon molecule that functions as the backbone of these membrane lipids. Within an individual glycerophospholipid, fatty acids are attached to the first and second carbons, and the phosphate group is attached to the third carbon of the glycerol backbone. Variable head groups are attached to the phosphate. Space-filling models of these molecules reveal their cylindrical shape, a geometry that allows glycerophospholipids to align side-by-side to form broad sheets (Figure 1).

Glycerophospholipids are by far the most abundant lipids in cell membranes. Like all lipids, they are insoluble in water, but their unique geometry causes them to aggregate into bilayers without any energy input. This is because they are two-faced molecules, with hydrophilic (water-loving) phosphate heads and hydrophobic (water-fearing) hydrocarbon tails of fatty acids. In water, these molecules spontaneously align — with their heads facing outward and their tails lining up in the bilayer's interior. Thus, the hydrophilic heads of the glycerophospholipids in a cell's plasma membrane face both the water-based cytoplasm and the exterior of the cell.

Altogether, lipids account for about half the mass of cell membranes. Cholesterol molecules, although less abundant than glycerophospholipids, account for about 20 percent of the lipids in animal cell plasma membranes. However, cholesterol is not present in bacterial membranes or mitochondrial membranes. Also, cholesterol helps regulate the stiffness of membranes, while other less prominent lipids play roles in cell signaling and cell recognition.

In addition to lipids, membranes are loaded with proteins. In fact, proteins account for roughly half the mass of most cellular membranes. Many of these proteins are embedded into the membrane and stick out on both sides; these are called transmembrane proteins. The portions of these proteins that are nested amid the hydrocarbon tails have hydrophobic surface characteristics, and the parts that stick out are hydrophilic .

At physiological temperatures, cell membranes are fluid; at cooler temperatures, they become gel-like. Scientists who model membrane structure and dynamics describe the membrane as a fluid mosaic in which transmembrane proteins can move laterally in the lipid bilayer. Therefore, the collection of lipids and proteins that make up a cellular membrane relies on natural biophysical properties to form and function. In living cells, however, many proteins are not free to move. They are often anchored in place within the membrane by tethers to proteins outside the cell, cytoskeletal elements inside the cell, or both.

What Do Membranes Do?

Cell membranes serve as barriers and gatekeepers. They are semi-permeable, which means that some molecules can diffuse across the lipid bilayer but others cannot. Small hydrophobic molecules and gases like oxygen and carbon dioxide cross membranes rapidly. Small polar molecules, such as water and ethanol, can also pass through membranes, but they do so more slowly. On the other hand, cell membranes restrict diffusion of highly charged molecules, such as ions, and large molecules, such as sugars and amino acids. The passage of these molecules relies on specific transport proteins embedded in the membrane.

Membrane transport proteins are specific and selective for the molecules they move, and they often use energy to catalyze passage. Also, these proteins transport some nutrients against the concentration gradient, which requires additional energy. The ability to maintain concentration gradients and sometimes move materials against them is vital to cell health and maintenance. Thanks to membrane barriers and transport proteins, the cell can accumulate nutrients in higher concentrations than exist in the environment and, conversely, dispose of waste products .

Other transmembrane proteins have communication-related jobs. These proteins bind signals, such as hormones or immune mediators, to their extracellular portions. Binding causes a conformational change in the protein that transmits a signal to intracellular messenger molecules. Like transport proteins, receptor proteins are specific and selective for the molecules they bind.

Peripheral membrane proteins are associated with the membrane but are not inserted into the bilayer. Rather, they are usually bound to other proteins in the membrane. Some peripheral proteins form a filamentous network just under the membrane that provides attachment sites for transmembrane proteins. Other peripheral proteins are secreted by the cell and form an extracellular matrix that functions in cell recognition.

How Diverse Are Cell Membranes?

In contrast to prokaryotes, eukaryotic cells have not only a plasma membrane that encases the entire cell, but also intracellular membranes that surround various organelles. In such cells, the plasma membrane is part of an extensive endomembrane system that includes the endoplasmic reticulum (ER), the nuclear membrane, the Golgi apparatus, and lysosomes. Membrane components are exchanged throughout the endomembrane system in an organized fashion. For instance, the membranes of the ER and the Golgi apparatus have different compositions, and the proteins that are found in these membranes contain sorting signals, which are like molecular zip codes that specify their final destination.

Mitochondria and chloroplasts are also surrounded by membranes, but they have unusual membrane structures — specifically, each of these organelles has two surrounding membranes instead of just one. The outer membrane of mitochondria and chloroplasts has pores that allow small molecules to pass easily. The inner membrane is loaded with the proteins that make up the electron transport chain and help generate energy for the cell. The double membrane enclosures of mitochondria and chloroplasts are similar to certain modern-day prokaryotes and are thought to reflect these organelles' evolutionary origins.

Conclusion

Membranes are made of lipids and proteins, and they serve a variety of barrier functions for cells and intracellular organelles. Membranes keep the outside "out" and the inside "in," allowing only certain molecules to cross and relaying messages via a chain of molecular events

Cytoskeletal Networks Provide Spatial Organization and Mechanical Support to Eukaryotic Cells

The cytoskeleton is a structure that helps cells maintain their shape and internal organization, and it also provides mechanical support that enables cells to carry out essential functions like division and movement. There is no single cytoskeletal component. Rather, several different components work together to form the cytoskeleton.

What Is the Cytoskeleton Made Of?

The cytoskeleton of eukaryotic cells is made of filamentous proteins, and it provides mechanical support to the cell and its cytoplasmic constituents. All cytoskeletons consist of three major classes of elements that differ in size and in protein composition. Microtubules are the largest type of filament, with a diameter of about 25 nanometers (nm), and they are composed of a protein called tubulin. Actin filaments are the smallest type, with a diameter of only about 6 nm, and they are made of a protein called actin. Intermediate filaments, as their name suggests, are mid-sized, with a diameter of about 10 nm. Unlike actin filaments and microtubules, intermediate filaments are constructed from a number of different subunit proteins.

What Do Microtubules Do?

Tubulin contains two polypeptide subunits, and dimers of these subunits string together to make long strands called protofilaments. Thirteen protofilaments then come together to form the hollow, straw-shaped filaments of microtubules. Microtubules are ever-changing, with reactions constantly adding and subtracting tubulin dimers at both ends of the filament (Figure 1). The rates of change at either end are not balanced — one end grows more rapidly and is called the plus end, whereas the other end is known as the minus end. In cells, the minus ends of microtubules are anchored in structures called microtubule organizing centers (MTOCs). The primary MTOC in a cell is called the centrosome, and it is usually located adjacent to the nucleus.

Microtubules tend to grow out from the centrosome to the plasma membrane. In nondividing cells, microtubule networks radiate out from the centrosome to provide the basic organization of the cytoplasm, including the positioning of organelles.

What Do Actin Filaments Do?

The protein actin is abundant in all eukaryotic cells. It was first discovered in skeletal muscle, where actin filaments slide along filaments of another protein called myosin to make the cells contract. (In nonmuscle cells, actin filaments are less organized and myosin is much less prominent.) Actin filaments are made up of identical actin proteins arranged in a long spiral chain. Like microtubules, actin filaments have plus and minus ends, with more ATP-powered growth occurring at a filament's plus end (Figure 2).

In many types of cells, networks of actin filaments are found beneath the cell cortex, which is the meshwork of membrane-associated proteins that supports and strengthens the plasma membrane. Such networks allow cells to hold — and move — specialized shapes, such as the brush border of microvilli. Actin filaments are also involved in cytokinesis and cell movement (Figure 3).

What Do Intermediate Filaments Do?

Intermediate filaments come in several types, but they are generally strong and ropelike. Their functions are primarily mechanical and, as a class, intermediate filaments are less dynamic than actin filaments or microtubules. Intermediate filaments commonly work in tandem with microtubules, providing strength and support for the fragile tubulin structures.

All cells have intermediate filaments, but the protein subunits of these structures vary. Some cells have multiple types of intermediate filaments, and some intermediate filaments are associated with specific cell types. For example, neurofilaments are found specifically in neurons (most prominently in the long axons of these cells), desmin filaments are found specifically in muscle cells, and keratins are found specifically in epithelial cells. Other intermediate filaments are distributed more widely. For example, vimentin filaments are found in a broad range of cell types and frequently colocalize with microtubules. Similarly, lamins are found in all cell types, where they form a meshwork that reinforces the inside of the nuclear membrane. Note that intermediate filaments are not polar in the way that actin or tubulin are (Figure 4).

How Do Cells Move?

Cytoskeletal filaments provide the basis for cell movement. For instance, cilia and (eukaryotic) flagella move as a result of microtubules sliding along each other. In fact, cross sections of these tail-like cellular extensions show organized arrays of microtubules.

Other cell movements, such as the pinching off of the cell membrane in the final step of cell division (also known as cytokinesis) are produced by the contractile capacity of actin filament networks. Actin filaments are extremely dynamic and can rapidly form and disassemble. In fact, this dynamic action underlies the crawling behavior of cells such as amoebae. At the leading edge of a moving cell, actin filaments are rapidly polymerizing; at its rear edge, they are quickly depolymerizing (Figure 5). A large number of other proteins participate in actin assembly and disassembly as well.

Conclusion

The cytoskeleton of a cell is made up of microtubules, actin filaments, and intermediate filaments. These structures give the cell its shape and help organize the cell's parts. In addition, they provide a basis for movement and cell division.

The Endoplasmic Reticulum, Golgi Apparatus, and Lysosomes Are Part of an Extensive Endomembrane System in Eukaryotic Cells

Cells have extensive sets of intracellular membranes, which together compose the endomembrane system. The endomembrane system was first discovered in the late 1800s when scientist Camillo Golgi noticed that a certain stain selectively marked only some internal cellular membranes. Golgi thought that these intracellular membranes were interconnected, but advances in microscopy and biochemical studies of the various membrane-encased organelles later made it clear the organelles in the endomembrane system are separate compartments with specific functions. These structures do exchange membrane material, however, via a special type of transport.

Today, scientists know that the endomembrane system includes the endoplasmic reticulum (ER), Golgi apparatus, and lysosomes. Vesicles also allow the exchange of membrane components with a cell's plasma membrane.

How Are Cell Membranes Synthesized?

Membranes and their constituent proteins are assembled in the ER. This organelle contains the enzymes involved in lipid synthesis, and as lipids are manufactured in the ER, they are inserted into the organelle's own membranes. This happens in part because the lipids are too hydrophobic to dissolve into the cytoplasm.

Similarly, transmembrane proteins have enough hydrophobic surfaces that they are also inserted into the ER membrane while they are still being synthesized. Here, future membrane proteins make their way to the ER membrane with the help of a signal sequence in the newly translated protein. The signal sequence stops translation and directs the ribosomes — which are carrying the unfinished proteins — to dock with ER proteins before finishing their work. Translation then recommences after the signal sequence docks with the ER, and it takes place within the ER membrane. Thus, by the time the protein achieves its final form, it is already inserted into a membrane (Figure 1).

The proteins that will be secreted by a cell are also directed to the ER during translation, where they end up in the lumen, the internal cavity, where they are then packaged for vesicular release from the cell. The hormones insulin and erythropoietin (EPO) are both examples of vesicular proteins.

How Are Organelle Membranes Maintained?

The ER, Golgi apparatus, and lysosomes are all members of a network of membranes, but they are not continuous with one another. Therefore, the membrane lipids and proteins that are synthesized in the ER must be transported through the network to their final destination in membrane-bound vesicles. Cargo-bearing vesicles pinch off of one set of membranes and travel along microtubule tracks to the next set of membranes, where they fuse with these structures. Trafficking occurs in both directions; the forward direction takes vesicles from the site of synthesis to the Golgi apparatus and next to a cell's lysosomes or plasma membrane. Vesicles that have released their cargo return via the reverse direction. The proteins that are synthesized in the ER have, as part of their amino acid sequence, a signal that directs them where to go, much like an address directs a letter to its destination.

Soluble proteins are carried in the lumens of vesicles. Any proteins that are destined for a lysosome are delivered to the lysosome interior when the vesicle that carries them fuses with the lysosomal membrane and joins its contents. In contrast, the proteins that will be secreted by a cell, such as insulin and EPO, are held in storage vesicles. When signaled by the cell, these vesicles fuse with the plasma membrane and release their contents into the extracellular space.

What Does the Golgi Apparatus Do?

The Golgi apparatus functions as a molecular assembly line in which membrane proteins undergo extensive post-translational modification. Many Golgi reactions involve the addition of sugar residues to membrane proteins and secreted proteins. The carbohydrates that the Golgi attaches to membrane proteins are often quite complex, and their synthesis requires multiple steps.

In electron micrographs, the Golgi apparatus looks like a set of flattened sacs. Vesicles that bud off from the ER fuse with the closest Golgi membranes, called the cis-Golgi. Molecules then travel through the Golgi apparatus via vesicle transport until they reach the end of the assembly line at the farthest sacs from the ER — called the trans-Golgi. At each workstation along the assembly line, Golgi enzymes catalyze distinct reactions. Later, as vesicles of membrane lipids and proteins bud off from the trans-Golgi, they are directed to their appropriate destinations — either lysosomes, storage vesicles, or the plasma membrane (Figure 2).

What Do Lysosomes Do?

Lysosomes break down macromolecules into their constituent parts, which are then recycled. These membrane-bound organelles contain a variety of enzymes called hydrolases that can digest proteins, nucleic acids, lipids, and complex sugars. The lumen of a lysosome is more acidic than the cytoplasm. This environment activates the hydrolases and confines their destructive work to the lysosome. In plants and fungi, lysosomes are called acidic vacuoles.

Lysosomes are formed by the fusion of vesicles that have budded off from the trans-Golgi. The sorting system recognizes address sequences in the hydrolytic enzymes and directs them to growing lysosomes. In addition, vesicles that bud off from the plasma membrane via endocytosis are also sent to lysosomes, where their contents — fluid and molecules from the extracellular environment — are processed. The process of endocytosis is an example of reverse vesicle trafficking, and it plays an important role in nutrition and immunity as well as membrane recycling. Lysosomes break down and thus disarm many kinds of foreign and potentially pathogenic materials that get into the cell through such extracellular sampling.

Conclusion

The endomembrane system of eukaryotic cells consists of the ER, the Golgi apparatus, and lysosomes. Membrane components, including proteins and lipids, are exchanged among these organelles and the plasma membrane via vesicular transport with the help of molecular tags that direct specific components to their proper destinations.

Mitochondria Are Independently Replicating Organelles That Supply Much of the Energy of the Cell

Mitochondria are unusual organelles. They act as the power plants of the cell, are surrounded by two membranes, and have their own genome. They also divide independently of the cell in which they reside, meaning mitochondrial replication is not coupled to cell division. Some of these features are holdovers from the ancient ancestors of mitochondria, which were likely free-living prokaryotes.

What Is the Origin of Mitochondria?

Mitochondria are thought to have originated from an ancient symbiosis that resulted when a nucleated cell engulfed an aerobic prokaryote. The engulfed cell came to rely on the protective environment of the host cell, and, conversely, the host cell came to rely on the engulfed prokaryote for energy production. Over time, the descendants of the engulfed prokaryote developed into mitochondria, and the work of these organelles — using oxygen to create energy — became critical to eukaryotic evolution.

Modern mitochondria have striking similarities to some modern prokaryotes, even though they have diverged significantly since the ancient symbiotic event. For example, the inner mitochondrial membrane contains electron transport proteins like the plasma membrane of prokaryotes, and

Concepts of Cell and Molecular Biology

mitochondria also have their own prokaryote-like circular genome. One difference is that these organelles are thought to have lost most of the genes once carried by their prokaryotic ancestor. Although present-day mitochondria do synthesize a few of their own proteins, the vast majority of the proteins they require are now encoded in the nuclear genome.

What Is the Purpose of a Mitochondrial Membranes?

As previously mentioned, mitochondria contain two major membranes. The outer mitochondrial membrane fully surrounds the inner membrane, with a small intermembrane space in between. The outer membrane has many protein-based pores that are big enough to allow the passage of ions and molecules as large as a small protein. In contrast, the inner membrane has much more restricted permeability, much like the plasma membrane of a cell. The inner membrane is also loaded with proteins involved in electron transport and ATP synthesis. This membrane surrounds the mitochondrial matrix, where the citric acid cycle produces the electrons that travel from one protein complex to the next in the inner membrane. At the end of this electron transport chain, the final electron acceptor is oxygen, and this ultimately forms water (H_2O). At the same time, the electron transport chain produces ATP. (This is why the the process is called oxidative phosphorylation.)

During electron transport, the participating protein complexes push protons from the matrix out to the intermembrane space. This creates a concentration gradient of protons that another protein complex, called ATP synthase, uses to power synthesis of the energy carrier molecule ATP (Figure 2).

Is the Mitochondrial Genome Still Functional?

Mitochondrial genomes are very small and show a great deal of variation as a result of divergent evolution. Mitochondrial genes that have been conserved across evolution include rRNA genes, tRNA genes, and a small number of genes that encode proteins involved in electron transport and ATP synthesis. The mitochondrial genome retains similarity to its prokaryotic ancestor, as does some of the machinery mitochondria use to synthesize proteins. In fact, mitochondrial rRNAs more closely resemble bacterial rRNAs than the eukaryotic rRNAs found in cell cytoplasm. In addition, some of the codons that mitochondria use to specify amino acids differ from the standard eukaryotic codons.

Still, the vast majority of mitochondrial proteins are synthesized from nuclear genes and transported into the mitochondria. These include the enzymes required for the citric acid cycle, the proteins involved in DNA replication and transcription, and ribosomal proteins. The protein complexes of the respiratory chain are a mixture of proteins encoded by mitochondrial genes and proteins encoded by nuclear genes. Proteins in both the outer and inner mitochondrial membranes help transport newly synthesized, unfolded proteins from the cytoplasm into the matrix, where folding ensues (Figure 3).

How Many Mitochondria Do Cells Have?

Mitochondria cannot be made "from scratch" because they need both mitochondrial and nuclear gene products. These organelles replicate by dividing in two, using a process similar to the simple, asexual form of cell division employed by bacteria. Video microscopy shows that mitochondria are incredibly dynamic. They are constantly dividing, fusing, and changing shape. Indeed, a single mitochondrion may contain multiple copies of its genome at any given time.

Logically, mitochondria multiply when a the energy needs of a cell increase. Therefore, power-hungry cells have more mitochondria than cells with lower energy needs. For example, repeatedly stimulating a muscle cell will spur the production of more mitochondria in that cell, to keep up with energy demand.

Conclusion

Mitochondria, the so-called "powerhouses" of cells, are unusual organelles in that they are surrounded by a double membrane and retain their own small genome. They also divide independently of the cell cycle by simple fission. Mitochondrial division is stimulated by energy demand, so cells with an increased need for energy contain greater numbers of these organelles than cells with lower energy needs.

Plant Cells Have Chloroplasts and Other Structures Not Present in Animal Cells

Plant cells have several structures not found in other eukaryotes. In particular, organelles called chloroplasts allow plants to capture the energy of the Sun in energy-rich molecules; cell walls allow plants to have rigid structures as varied as wood trunks and supple leaves; and vacuoles allow plant cells to change size.

What Is the Origin of Chloroplasts?

Like mitochondria, chloroplasts likely originated from an ancient symbiosis, in this case when a nucleated cell engulfed a photosynthetic prokaryote. Indeed, chloroplasts resemble modern cyanobacteria, which remain similar to the cyanobacteria of 3 million years ago. However, the evolution of photosynthesis goes back even further, to the earliest cells that evolved the ability to capture light energy and use it to produce energy-rich molecules. When these organisms developed the ability to split water molecules and use the electrons from these molecules, photosynthetic cells started generating oxygen — an event that had dramatic consequences for the evolution of all living things on Earth.

Today, chloroplasts retain small, circular genomes that resemble those of cyanobacteria, although they are much smaller. (Mitochondrial genomes are even smaller than the genomes of chloroplasts.) Coding sequences for the majority of chloroplast proteins have been lost, so these proteins are now encoded by the nuclear genome, synthesized in the cytoplasm, and transported from the cytoplasm into the chloroplast.

What Is the Function of Chloroplast Membranes?

Like mitochondria, chloroplasts are surrounded by two membranes. The outer membrane is permeable to small organic molecules, whereas the inner membrane is less permeable and studded with transport proteins. The innermost matrix of chloroplasts, called the stroma, contains metabolic enzymes and multiple copies of the chloroplast genome.

Chloroplasts also have a third internal membrane called the thylakoid membrane, which is extensively folded and appears as stacks of flattened disks in electron micrographs. The thylakoids contain the light-harvesting complex, including pigments such as chlorophyll, as well as the electron transport chains used in photosynthesis.

What Is the Cell Wall?

Besides the presence of chloroplasts, another major difference between plant and animal cells is the presence of a cell wall. The cell wall surrounds the plasma membrane of plant cells and provides tensile strength and protection against mechanical and osmotic stress. It also allows cells to develop turgor pressure, which is the pressure of the cell contents against the cell wall. Plant cells have high concentrations of molecules dissolved in their cytoplasm, which causes water to come into the cell under normal conditions and makes the cell's central vacuole swell and press against the cell wall. With a healthy supply of water, turgor pressure keeps a plant from wilting. In drought, a plant may wilt, but its cell walls help maintain the structural integrity of its stems, leaves, and other structures, despite a shrinking, less turgid vacuole.

Plant cell walls are primarily made of cellulose, which is the most abundant macromolecule on Earth. Cellulose fibers are long, linear polymers of hundreds of glucose molecules. These fibers aggregate into bundles of about 40, which are called microfibrils. Microfibrils are embedded in a hydrated network of other polysaccharides. The cell wall is assembled in place. Precursor components are synthesized inside the cell and then assembled by enzymes associated with the cell membrane (Figure 3).

What Are Vacuoles?

Plant cells additionally possess large, fluid-filled vesicles called vacuoles within their cytoplasm. Vacuoles typically compose about 30 percent of a cell's volume, but they can fill as much as 90 percent of the intracellular space. Plant cells use vacuoles to adjust their size and turgor pressure. Vacuoles usually account for changes in cell size when the cytoplasmic volume stays constant.

Some vacuoles have specialized functions, and plant cells can have more than one type of vacuole. Vacuoles are related to lysosomes and share some functions with these structures; for instance, both contain degradative enzymes for breaking down macromolecules. Vacuoles can also serve as storage compartments for nutrients and metabolites. For instance, proteins are stored in the vacuoles of seeds, and rubber and opium are metabolites that are stored in plant vacuoles.

Conclusion

Plant cells have certain distinguishing features, including chloroplasts, cell walls, and intracellular vacuoles. Photosynthesis takes place in chloroplasts; cell walls allow plants to have strong, upright structures; and vacuoles help regulate how cells handle water and storage of other molecules.

How Do Cells Sense Their Environment?

Cells may be self-sustaining units of life, but they don't live in isolation. Their survival depends on receiving and processing information from the outside environment, whether that information pertains to the availability of nutrients, changes in temperature, or variations in light levels.

Cells also can communicate with one another — and change their own internal workings in response — by way of a variety of chemical and mechanical signals. In multicellular organisms, cell signaling allows for specialization of groups of cells. Multiple cell types can then join together to form tissues, such as muscle, blood, and brain tissue. In single-celled organisms, signaling allows populations of cells to coordinate with one another and work as a team to accomplish tasks no single cell could carry out on its own.

Cells Receive and Process a Diverse Set of Chemical Signals and Sensory Stimuli

In order to respond to changes in their immediate environment, cells must be able to receive and process signals that originate outside their borders. Individual cells often receive many signals simultaneously, and they then integrate the information they receive into a unified action plan. But cells aren't just targets. They also send out messages to other cells both near and far.

What Kind of Signals Do Cells Receive?

Most cell signals are chemical in nature. For example, prokaryotic organisms have sensors that detect nutrients and help them navigate toward food sources. In multicellular organisms, growth factors, hormones, neurotransmitters, and extracellular matrix components are some of the many types of chemical signals cells use. These substances can exert their effects locally, or they might travel over long distances. For instance, neurotransmitters are a class of short-range signaling molecules that travel across the tiny spaces between adjacent neurons or between neurons and muscle cells. Other signaling molecules must move much farther to reach their targets. One example is follicle-stimulating hormone, which travels from the mammalian brain to the ovary, where it triggers egg release.

Some cells also respond to mechanical stimuli. For example, sensory cells in the skin respond to the pressure of touch, whereas similar cells in the ear react to the movement of sound waves. In addition, specialized cells in the human vascular system detect changes in blood pressure — information that the body uses to maintain a consistent cardiac load.

How Do Cells Recognize Signals?

Cells have proteins called receptors that bind to signaling molecules and initiate a physiological response. Different receptors are specific for different molecules. Dopamine receptors bind dopamine, insulin receptors bind insulin, nerve growth factor receptors bind nerve growth factor, and so on. In fact, there are hundreds of receptor types found in cells, and varying cell types have different populations of receptors. Receptors can also respond directly to light or pressure, which makes cells sensitive to events in the atmosphere.

Receptors are generally transmembrane proteins, which bind to signaling molecules outside the cell and subsequently transmit the signal through a sequence of molecular switches to internal signaling pathways. Membrane receptors fall into three major classes: G-protein-coupled receptors, ion channel receptors, and enzyme-linked receptors. The names of these receptor classes refer to the mechanism by which the receptors transform external signals into internal ones — via protein action, ion channel opening, or enzyme activation, respectively. Because membrane receptors interact with both extracellular signals and molecules within the cell, they permit signaling molecules to affect cell function without actually entering the cell. This is important because most signaling molecules are either too big or too charged to cross a cell's plasma membrane (Figure 1).

Not all receptors exist on the exterior of the cell. Some exist deep inside the cell, or even in the nucleus. These receptors typically bind to molecules that can pass through the plasma membrane, such as gases like nitrous oxide and steroid hormones like estrogen.

How Do Cells Respond to Signals?

Once a receptor protein receives a signal, it undergoes a conformational change, which in turn launches a series of biochemical reactions within the cell. These intracellular signaling pathways, also called signal transduction cascades, typically amplify the message, producing multiple intracellular signals for every one receptor that is bound.

Activation of receptors can trigger the synthesis of small molecules called second messengers, which initiate and coordinate intracellular signaling pathways. For example, cyclic AMP (cAMP) is a common second messenger involved in signal transduction cascades. (In fact, it was the first second messenger ever discovered.) cAMP is synthesized from ATP by the enzyme adenylyl cyclase, which resides in the cell membrane. The activation of adenylyl cyclase can result in the manufacture of hundreds or even thousands of cAMP molecules. These cAMP molecules activate the enzyme protein kinase A (PKA), which then phosphorylates multiple protein substrates by attaching phosphate groups to them. Each step in the cascade further amplifies the initial signal, and the phosphorylation reactions mediate both short- and long-term responses in the cell (Figure 2). How does cAMP stop signaling? It is degraded by the enzyme phosphodiesterase.

Other examples of second messengers include diacylglycerol (DAG) and inositol 1,4,5-triphosphate (IP3), which are both produced by the enzyme phospholipase, also a membrane protein. IP3 causes the release of Ca^{2+} — yet another second messenger — from intracellular stores. Together, DAG and Ca^{2+} activate another enzyme called protein kinase C (PKC).

How Do Signals Affect Cell Function?

Protein kinases such as PKA and PKC catalyze the transfer of phosphate groups from ATP molecules to protein molecules. Within proteins, the amino acids serine, threonine, and tyrosine are especially common sites for phosphorylation. These phosphorylation reactions control the activity of many enzymes involved in intracellular signaling pathways. Specifically, the addition of phosphate groups causes a conformational change in the enzymes, which can either activate or inhibit the enzyme activity. Then, when appropriate, protein phosphatases remove the phosphate groups from the enzymes, thereby reversing the effect on enzymatic activity.

Phosphorylation allows for intricate control of protein function. Phosphate groups can be added to multiple sites in a single protein, and a single protein may in turn be the substrate for multiple kinases and phosphatases.

At any one time, a cell is receiving and responding to numerous signals, and multiple signal transduction pathways are operating in its cytoplasm. Many points of intersection exist among these pathways. For instance, a single second messenger or protein kinase might play a role in more than one pathway. Through this network of signaling pathways, the cell is constantly integrating all the information it receives from its external environment.

Conclusion

Cells typically receive signals in chemical form via various signaling molecules. When a signaling molecule joins with an appropriate receptor on a cell surface, this binding triggers a chain of events that not only carries the signal to the cell interior, but amplifies it as well. Cells can also send signaling molecules to other cells. Some of these chemical signals — including neurotransmitters — travel only a short distance, but others must go much farther to reach their targets.

G-Protein-Coupled Receptors Play Many Different Roles in Eukaryotic Cell Signaling

G-protein-coupled receptors (GPCRs) are the largest and most diverse group of membrane receptors in eukaryotes. These cell surface receptors act like an inbox for messages in the form of light energy, peptides, lipids, sugars, and proteins. Such messages inform cells about the presence or absence of life-sustaining light or nutrients in their environment, or they convey information sent by other cells.

GPCRs play a role in an incredible array of functions in the human body, and increased understanding of these receptors has greatly affected modern medicine. In fact, researchers estimate that between one-third and one-half of all marketed drugs act by binding to GPCRs.

What Do GPCRs Look Like?

GPCRs bind a tremendous variety of signaling molecules, yet they share a common architecture that has been conserved over the course of evolution. Many present-day eukaryotes — including animals, plants, fungi, and protozoa — rely on these receptors to receive information from their environment. For example, simple eukaryotes such as yeast have GPCRs that sense glucose and mating factors. Not surprisingly, GPCRs are involved in considerably more functions in multicellular organisms. Humans alone have nearly 1,000 different GPCRs, and each one is highly specific to a particular signal.

GPCRs consist of a single polypeptide that is folded into a globular shape and embedded in a cell's plasma membrane. Seven segments of this molecule span the entire width of the membrane — explaining why GPCRs are sometimes called seven-transmembrane receptors — and the intervening portions loop both inside and outside the cell. The extracellular loops form part of the pockets at which signaling molecules bind to the GPCR.

What Do GPCRs Do?

As their name implies, GPCRs interact with G proteins in the plasma membrane. When an external signaling molecule binds to a GPCR, it causes a conformational change in the GPCR. This change then triggers the interaction between the GPCR and a nearby G protein.

G proteins are specialized proteins with the ability to bind the nucleotides guanosine triphosphate (GTP) and guanosine diphosphate (GDP). Some G proteins, such as the signaling protein Ras, are small proteins with a single subunit. However, the G proteins that associate with GPCRs are heterotrimeric, meaning they have three different subunits: an alpha subunit, a beta subunit, and a gamma subunit. Two of these subunits — alpha and gamma — are attached to the plasma membrane by lipid anchors.

A G protein alpha subunit binds either GTP or GDP depending on whether the protein is active (GTP) or inactive (GDP). In the absence of a signal, GDP attaches to the alpha subunit, and the entire G protein-GDP complex binds to a nearby GPCR. This arrangement persists until a signaling molecule joins with the GPCR. At this point, a change in the conformation of the GPCR activates the G protein, and GTP physically replaces the GDP bound to the alpha subunit. As a result, the G protein subunits dissociate into two parts: the GTP-bound alpha subunit and a beta-gamma dimer. Both parts remain anchored to the plasma membrane, but they are no longer bound to the GPCR, so they can now diffuse laterally to interact with other membrane proteins. G proteins remain active as long as their alpha subunits are joined with GTP. However, when this GTP is hydrolyzed back to GDP, the subunits once again assume the form of an inactive heterotrimer, and the entire G protein reassociates with the now-inactive GPCR. In this way, G proteins work like a switch — turned on or off by signal-receptor interactions on the cell's surface.

Whenever a G protein is active, both its GTP-bound alpha subunit and its beta-gamma dimer can relay messages in the cell by interacting with other membrane proteins involved in signal transduction. Specific targets for activated G proteins include various enzymes that produce second messengers, as well as certain ion channels that allow ions to act as second messengers. Some G proteins stimulate the activity of these targets, whereas others are inhibitory. Vertebrate genomes contain multiple genes that encode the alpha, beta, and gamma subunits of G proteins. The many different subunits encoded by these genes combine in multiple ways to produce a diverse family of G proteins (Figure 2).

What Second Messengers Do GPCR Signals Trigger in Cells?

Activation of a single G protein can affect the production of hundreds or even thousands of second messenger molecules (Figure 3). (Recall that second messengers — such as cyclic AMP [cAMP], diacylglycerol [DAG], and inositol 1, 4, 5-triphosphate [IP3] — are small molecules that initiate and coordinate intracellular signaling pathways.) One especially common target of activated G proteins is adenylyl cyclase, a membrane-associated enzyme that, when activated by the GTP-bound alpha subunit, catalyzes synthesis of the second messenger cAMP from molecules of ATP. In humans, cAMP is involved in responses to sensory input, hormones, and nerve transmission, among others.

Phospholipase C is another common target of activated G proteins. This membrane-associated enzyme catalyzes the synthesis of not one, but two second messengers — DAG and IP3 — from the membrane lipid phosphatidyl inositol. This particular pathway is critical to a wide variety of human bodily processes. For instance, thrombin receptors in platelets use this pathway to promote blood clotting.

Conclusion

GPCRs are a large family of cell surface receptors that respond to a variety of external signals. Binding of a signaling molecule to a GPCR results in G protein activation, which in turn triggers the production of any number of second messengers. Through this sequence of events, GPCRs help regulate an incredible range of bodily functions, from sensation to growth to hormone responses.

Ion Channel Receptors Generate Electrical Signals in Response to Chemical Signals

Certain cells, commonly called excitable cells, are unique because of their ability to generate electrical signals. Although several types of excitable cells exist — including neurons, muscle cells, and touch receptor cells — all of them use ion channel receptors to convert chemical or mechanical messages into electrical signals.

Like all cells, an excitable cell maintains a different concentration of ions in its cytoplasm than exists in its extracellular environment. Together, these concentration differences create a small electrical potential across the plasma membrane. Then, when conditions are right, specialized channels in the plasma membrane open and allow rapid ion movement into or out of the cell, and this movement creates an electrical signal. But what do these channels look like, and how do they function? Also, how do the electrical signals generated by excitable cells differ from the other types of signals involved in cellular communication?

What Are Ion Channel Receptors?

Ion channel receptors are usually multimeric proteins located in the plasma membrane. Each of these proteins arranges itself so that it forms a passageway or pore extending from one side of the membrane to the other. These passageways, or ion channels, have the ability to open and close in response to chemical or mechanical signals. When an ion channel is open, ions move into or out of the cell in single-file fashion. Individual ion channels are specific to particular ions, meaning that they usually allow only a single type of ion to pass through them. Both the amino acids that line a channel and the physical width of the channel determine which ions are able to wiggle through from the cell exterior to its interior, and vice versa. The opening of an ion channel is a fleeting event. Within a few milliseconds of opening, most ion channels close and enter a resting state, where they are unresponsive to signals for a short period of time.

How Are Electrical Signals Propagated?

The opening of ion channels alters the charge distribution across the plasma membrane. Recall that the ionic composition of the cytoplasm is quite different from that of the extracellular environment. For instance, the concentration of sodium ions in the cytoplasm is far lower than that in the cell's exterior environment. Conversely, potassium ions exist at higher concentrations within a cell than outside it. Such differences create a so-called electrochemical gradient, which is a combination of a chemical gradient and a chargegradient. The opening of ion channels permits the ions on either side of the plasma membrane to flow down this dual gradient. The exact direction of flow varies by ion type, and it depends on both the concentration difference and the voltage difference for each variety of ion. This ion flow results in the production of an electrical signal. The actual number of ions required to change the voltage across the membrane is quite small. During the short times that an ion channel is open, the concentration of a particular ion in the cytoplasm as a whole does not change significantly, only the concentration in the immediate vicinity of the channel. In excitable cells, the electrical signal initiated by ion channel receptor activity travels rapidly over the surface of the cell due to the opening of other ion channels that are sensitive to the voltage change caused by the initial channel opening.

Electrical signals travel much more rapidly than chemical signals, which depend on the process of molecular diffusion. As a consequence, excitable cells respond to signals much more rapidly than cells that rely solely on chemical signals (Figure 2). In fact, an electrical signal can traverse the entire length of a human nerve cell — a distance of as much as one meter — within only milliseconds.

How Do Different Types of Excitable Cells Work?

Neurons, muscle cells, and touch receptor cells are all excitable cells — which means they all have the capacity to transmit electrical signals. Each of these cells also has ion channel receptors clustered on a particular part of its surface. For example, the receptors that respond to chemical signals are generally located at synapses — or points of near contact between adjacent cells.

Of the various types of excitable cells that respond to chemical signals, neurons are perhaps the most familiar. When electrical signals reach the end of neurons, they trigger the release of chemical messengers called neurotransmitters. Each neurotransmitter then diffuses from its point of release on one side of the synapse to the cell on the other side of the synapse. If the neurotransmitter binds to an ion channel receptor on the target cell, the related ion channel opens, and an electrical signal propagates itself along the length of the target cell.

Neurons have ion channel receptors specific to many kinds of neurotransmitters. Some of these neurotransmitters act in an excitatory capacity, bringing their target cells ever closer to signal propagation. Other neurotransmitters exert an inhibitory effect, counteracting any excitatory input and lessening the chance that the target cell will fire.

Skeletal muscle cells also rely on chemical signals in order to generate electrical signals. These cells have synapses that are packed with receptors for acetylcholine, which is the primary neurotransmitter released by motor neurons. When acetylcholine binds to the receptors on a skeletal muscle cell, ion channels in that cell open, and this launches a sequence of events that results in contraction of the cell.

In contrast to neurons and skeletal muscle cells, some excitable cells have ion channels that open in response to mechanical stimuli rather than chemical signals. These include the hair cells of the mammalian inner ear and the touch receptor cells of both human finger pads and Venus fly traps. Cells that respond to touch have their ion channel receptors clustered at the position where contact usually occurs.

Conclusion

Excitable cells, such as fast-acting neurons and muscle cells, have specialized channels that open in response to a signal and permit rapid ion movement across the cell membrane. The opening of just a single ion channel alters the electrical charge on both sides of the membrane. The resulting charge differential then causes adjacent voltage-sensitive channels to open in chain-reaction fashion, creating a self-propagating electrical signal that travels down the entire length of the cell. Sometimes, this sequence of events is triggered when a chemical signal — such as a neurotransmitter — binds to an ion channel receptor on cell's surface. Other times, a cell's ion channels open in response to mechanical (rather than chemical) stimuli.

Receptor Tyrosine Kinases Regulate Cell Growth, Differentiation, and Survival

Although all cell membrane receptors receive and transmit signals from the environment, some of these receptors also double as enzymes. In such cases, the binding of a signaling molecule to the membrane receptor activates the receptor's inherent enzymatic activity. Of the various receptors that exhibit this capability, receptor tyrosine kinases (RTKs) make up the largest class. These cell surface receptors bind and respond to growth factors and other locally released proteins that are present at low concentrations. RTKs play important roles in the regulation of cell growth, differentiation, and survival.

When signaling molecules bind to RTKs, they cause neighboring RTKs to associate with each other, forming cross-linked dimers. Cross-linking activates the tyrosine kinase activity in these RTKs through phosphorylation — specifically, each RTK in the dimer phosphorylates multiple tyrosines on the other RTK. This process is called cross-phosphorylation.

What Do RTKs Look Like?

Once cross-phosphorylated, the cytoplasmic tails of RTKs serve as docking platforms for various intracellular proteins involved in signal transduction. These proteins have a particular domain — called SH2 — that binds to phosphorylated tyrosines in the cytoplasmic RTK receptor tails. More than one SH2-containing protein can bind at the same time to an activated RTK, allowing simultaneous activation of multiple intracellular signaling pathways. Ultimately, RTK activation brings about changes in gene transcription. Signaling becomes complex as signals travel from the membrane to the nucleus, due to crosstalk between intermediates in various signaling pathways in the cell (Figure 1).

One of the most common intracellular signaling pathways triggered by RTKs is known as the mitogen-activated protein (MAP) kinase cascade, because it involves three serine-threonine kinases. The pathway starts with the activation of Ras, a small G protein anchored to the plasma membrane. In its inactive state, Ras is bound to GDP. However, when SH2-containing proteins join with activated RTKs, they cause Ras to bind GTP in place of GDP and become active. Next, the GTP-bound Ras (which is not itself a kinase) activates the first serine-threonine kinase in the MAP kinase cascade. Each of the three kinases in this cascade then activates the next by phosphorylating it. Because all three kinases in this pathway phosphorylate multiple substrates, the initial signal is amplified at each step. Then, the final enzyme in the pathway phosphorylates transcription regulators, leading to a change in gene transcription. Many growth factors, including nerve growth factor and platelet-derived growth factor, use this pathway.

Not all RTKs use the MAP kinase cascade to send information to the nucleus. For example, insulin-like growth factor receptors activate the enzyme phosphoinositide 3-kinase, which phosphorylates inositol phospholipids in the cell membrane, leading in turn to a protein kinase cascade (distinct from the MAP kinase cascade) that relays the signal to the nucleus. Other RTKs use a more direct route to the nucleus. Transcriptional regulators known as STAT proteins, an acronym for signal transducers and activators of transcription, bind to the phosphorylated tyrosines in the receptors for cytokines and some hormones. Once activated, STAT proteins move directly into the nucleus, causing changes in transcription.

How Do RTK Signals Regulate Cells?

Cells possess many different RTKs that bind to a diverse set of extracellular signaling molecules, many of which are locally produced and present in low concentrations. These local cell-to-cell interactions are important for developing and maintaining the spatial orientation of tissues, which is crucial for higher-level functioning.

Growth factors and hormones are two especially important categories of signaling molecules that bind to RTKs. These molecules direct cell differentiation by determining patterns of gene transcription. Extracellular matrix proteins and certain surface proteins on neighboring cells can also bind to and activate RTKs. For example, upon binding to RTKs, surface proteins called ephrins help guide developmental processes involved in tissue architecture, final placement of nerve endings, and blood vessel maturation.

When RTKs don't function properly, cell growth and differentiation go awry. For instance, many cancers appear to involve mutations in RTKs. For this reason, RTKs are the targets of various drugs used in cancer chemotherapy. For example, the breast cancer drug Herceptin is an antibody that binds to and inhibits ErbB-2 — an RTK that is overexpressed in many metastatic breast cancers.

Conclusion

RTKs are transmembrane protein receptors that help cells interact with their neighbors in a tissue. RTKs differ from other cell surface receptors in that they contain intrinsic enzyme activity. In particular, the binding of a signaling molecule with an RTK activates tyrosine kinase in the cytoplasmic tail of the receptor. This activity then launches a series of enzymatic reactions that carry the signal to the nucleus, where it alters patterns of protein transcription.

Cells Sense the Presence of Other Cells and Their Environment

All cells rely on cell signaling to detect and respond to cues in their environment. This process not only promotes the proper functioning of individual cells, but it also allows communication and coordination among groups of cells — including the cells that make up organized communities called tissues. Because of cell signaling, tissues have the ability to carry out tasks no single cell could accomplish on its own.

Different types of tissues, such as bone, brain, and the lining of the gut, have characteristic features related to the number and types of cells they contain. Cell spacing is also critical to tissue function, so this geometry is precisely regulated. To preserve proper tissue architecture, adhesive molecules help maintain contact between nearby cells and structures, and tiny tunnel-like junctions allow the passage of ions and small molecules between adjacent cells. Meanwhile, signaling molecules relay positional information among the cells in a tissue, as well as between these cells and the extracellular matrix. These signaling pathways are critical to maintaining the state of equilibrium known as homeostasis within a tissue. For example, the processes involved in wound healing depend on positional information in order for normal tissue architecture to be restored. Such positional signals are also crucial for the development of adult structures in multicellular organisms. As tissues develop, clumps of unorganized cells grow and sort themselves according to signals they send and receive.

How Do Integrins Promote Tissue Structure and Function?

Within tissues, adhesive molecules allow cells to maintain contact with one another and with structures in the extracellular matrix. One especially important class of adhesive molecules is the integrins. Integrins are more than just mechanical links, however: They also relay signals both to and from cells. In this way, integrins play an important role in sensing the environment and controlling cell shape and motility.

Integrins are a diverse family of transmembrane proteins found in all animal cells. Even simple animals like sponges have these proteins. Each individual integrin consists of two main parts: an alpha subunit and a beta subunit. Variation in the alpha and beta subunits accounts for the wide variety of integrins observed throughout the animal kingdom. For example, humans alone have over 20 different kinds of integrins.

Integrins link the actin cytoskeleton of a cell to various external structures. The cytoplasmic portion of each integrin molecule binds to adaptor proteins that connect to the actin filaments inside the cell. The extracellular portion of the integrin then binds to molecules in the extracellular matrix or on the surface of other cells. Integrin attachments to neighboring cells can break and reform as a cell moves (Figure 1).

How Else Do the Cells within a Tissue Stay in Contact?

Beyond integrins, cells rely on several other adhesive proteins to maintain physical contact. As an example, consider the epithelial cells that line the inner and outer surfaces of the human body — including the skin, intestines, airway, and reproductive tract. These cells provide a dramatic example of the different kinds of cell-to-cell junctions, but the same junctions also exist in a wide range of other tissues.

The side surfaces of epithelial cells are tightly linked to those of neighboring cells, forming a sheet that acts as a barrier. Within this sheet, each individual cell has a set orientation. Through integrins, the basal end of each cell connects to a specialized layer of extracellular matrix called the basal lamina. In contrast, the apical end of each cell faces out into the environment — such as the inner cavity or lumen of the gut.

The side-to-side junctions that link epithelial cells are diverse in their protein makeup and function. The adhesive transmembrane proteins anchoring these junctions have extracellular portions that interact with similar proteins on adjacent cells. Protein complexes within each cell further connect the transmembrane adhesive proteins to the cytoskeleton. In particular, adaptor complexes bind adherens junctions to cytoskeletal actin, and other adaptor complexes bind desmosomes to intermediate filaments. Both of these types of junctional complexes provide cells and tissues with mechanical support, and they additionally recruit intracellular signaling molecules to relay positional information to the nucleus.

The lateral surfaces of epithelial cells also contain several other types of specialized junctions. Tight junctions form a seal between cells that is so strong that not even ions can pass across it. Gap junctions are involved in cellular communication — not just in epithelial tissue, but in other tissue types as well. Gap junctions are specialized connections that form a narrow pore between adjacent cells. These pores permit small molecules and ions to move from one cell to another. In this way, gap junctions provide metabolic and electrical coupling between cells. For example, cardiac tissue has extensive gap junctions, and the rapid movement of ions through these junctions helps the tissue beat in rhythm. Gap junctions may also open and close in response to metabolic signals (Figure 2, Figure 3).

Cell Death Can Be Prompted by a Signal

Cell signaling isn't just central to tissue architecture and function: It also plays an important role in the balance between cell growth and death. Although it sounds like a bad thing, apoptosis — or the process of programmed cell death — is an essential aspect of development. Without it, repair and replenishment processes would overrun tissues with new cells. The orderly demise of a certain proportion of cells is therefore necessary for normal tissue turnover and maintenance of homeostasis. Apoptosis is distinct from necrosis, a messier form of cell death that causes cells to literally swell and burst. Necrotic cell death is not programmed; rather, it occurs in response to trauma or injury.

A range of extracellular and intracellular signals can trigger either cell growth or apoptosis. When cells receive these signals from their neighbors or from other aspects of the external environment, they carefully weigh them against each other before choosing a course of action. For instance, signals that indicate a lack of nutrients or the presence of toxins would likely stall cell growth and promote apoptosis. Within the cell, damage to the DNA or loss of mitochondrial integrity might also result in programmed cell death.

Cells self-destruct cleanly and quickly during apoptosis, thanks to the activation of a variety of enzymes — proteases and nucleases — that break down proteins and nucleic acids, respectively. In fact, scientists look for a characteristic pattern of fragmentation and nuclear condensation within tissues as evidence that apoptosis has occurred.

Conclusion

Some cell signaling occurs on a local level, such as when cells interact with the surrounding extracellular matrix or with their immediate neighbors. This type of signaling is especially important to the structure and function of tissues. Various signaling molecules allow the cells within a tissue to share information about internal and external conditions. This information helps the cells arrange themselves, coordinate their functions, and even know when to grow and when to die. Some of these signaling molecules also function in an adhesive capacity — not just relaying messages between the cells in a tissue, but physically joining these cells to one another.

How Do Cells Know When to Divide?

Cells can replicate themselves. The ability to reproduce is part of what defines cells as living things. This single characteristic also helps explain many other phenomena of life as we know it, including the emergence of multicellular organisms, the wide variety of tissues observed in living things, and even the scourge of cancer.

The process by which a single cell divides into two daughter cells is called mitosis. Mitosis is an important part of a cell's life cycle — but the rest of this cycle, collectively known as interphase, is hardly static. During interphase, the cell carries out the everyday biochemical reactions associated with metabolism, and it also engages in several processes that will guide it through the next round of division. In addition, throughout the cell cycle there are multiple monitoring systems and checkpoints that help the cell determine if and when it should divide, whether it's time to advance to the next phase, or whether it's time to die and make room for a younger, healthier cell.

The various checks on cell growth that occur during interphase allow tissues to revitalize themselves without increasing in size. When these restraints fail, the results — including the growth and spread of cancer — can be devastating.

The Eukaryotic Cell Cycle Consists of Discrete Phases

The cellular life cycle, also called the cell cycle, includes many processes necessary for successful self-replication. Beyond carrying out the tasks of routine metabolism, the cell must duplicate its components — most importantly, its genome — so that it can physically split into two complete daughter cells. The cell must also pass through a series of checkpoints that ensure conditions are favorable for division.

What Phases Make Up the Eukaryotic Cell Cycle?

In eukaryotes, the cell cycle consists of four discrete phases: G1, S, G2, and M. The S or synthesis phase is when DNA replication occurs, and the M or mitosis phase is when the cell actually divides. The other two phases — G1 and G2, the so-called gap phases — are less dramatic but equally important. During G1, the cell conducts a series of checks before entering the S phase. Later, during G2, the cell similarly checks its readiness to proceed to mitosis.

Together, the G1, S, and G2 phases make up the period known as interphase. Cells typically spend far more time in interphase than they do in mitosis. Of the four phases, G1 is most variable in terms of duration, although it is often the longest portion of the cell cycle (Figure 1).

How Do Cells Monitor Their Progress through the Cell Cycle?

In order to move from one phase of its life cycle to the next, a cell must pass through numerous checkpoints. At each checkpoint, specialized proteins determine whether the necessary conditions exist. If so, the cell is free to enter the next phase. If not, progression through the cell cycle is halted. Errors in these checkpoints can have catastrophic consequences, including cell death or the unrestrained growth that is cancer.

Each part of the cell cycle features its own unique checkpoints. For example, during G1, the cell passes through a critical checkpoint that ensures environmental conditions (including signals from other cells) are favorable for replication. If conditions are not favorable, the cell may enter a resting state known as G0. Some cells remain in G0 for the entire lifetime of the organism in which they reside. For instance, the neurons and skeletal muscle cells of mammals are typically in G0.

Another important checkpoint takes place later in the cell cycle, just before a cell moves from G2 to mitosis. Here, a number of proteins scrutinize the cell's DNA, making sure it is structurally intact and properly replicated. The cell may pause at this point to allow time for DNA repair, if necessary.

Yet another critical cell cycle checkpoint takes place mid-mitosis. This check determines whether the chromosomes in the cell have properly attached to the spindle, or the network of microtubules that will separate them during cell division. This step decreases the possibility that the resulting daughter cells will have unbalanced numbers of chromosomes — a condition called aneuploidy.

How Do Scientists Study the Cell Cycle?

The cell cycle and its system of checkpoint controls show strong evolutionary conservation. As a result, all eukaryotes — from single-celled yeast to complex multicellular vertebrates — pass through the same four phases and same key checkpoints. This universality of the cell cycle and its checkpoint controls allows scientists to use relatively simple model organisms to learn more about cell division in eukaryotes of all types — including humans. In fact, two of the three scientists who received Nobel Prizes for cell cycle research used yeast as the subject of their investigations.

Conclusion

The eukaryotic cell cycle includes four phases necessary for proper growth and division. As a cell moves through each phase, it also passes through several checkpoints. These checkpoints ensure that mitosis occurs only when environmental conditions are favorable and the cellular genome has been precisely replicated. Collectively, this set of checks on division helps prevent chromosomal imbalance in newly produced daughter cells.

The Eukaryotic Cell Cycle Consists of Discrete Phases

The cellular life cycle, also called the cell cycle, includes many processes necessary for successful self-replication. Beyond carrying out the tasks of routine metabolism, the cell must duplicate its components — most importantly, its genome — so that it can physically split into two complete daughter cells. The cell must also pass through a series of checkpoints that ensure conditions are favorable for division.

What Phases Make Up the Eukaryotic Cell Cycle?

In eukaryotes, the cell cycle consists of four discrete phases: G1, S, G2, and M. The S or synthesis phase is when DNA replication occurs, and the M or mitosis phase is when the cell actually divides. The other two phases — G1 and G2, the so-called gap phases — are less dramatic but equally important. During G1, the cell conducts a series of checks before entering the S phase. Later, during G2, the cell similarly checks its readiness to proceed to mitosis.

Together, the G1, S, and G2 phases make up the period known as interphase. Cells typically spend far more time in interphase than they do in mitosis. Of the four phases, G1 is most variable in terms of duration, although it is often the longest portion of the cell cycle (Figure 1).

How Do Cells Monitor Their Progress through the Cell Cycle?

In order to move from one phase of its life cycle to the next, a cell must pass through numerous checkpoints. At each checkpoint, specialized proteins determine whether the necessary conditions exist. If so, the cell is free to enter the next phase. If not, progression through the cell cycle is halted. Errors in these checkpoints can have catastrophic consequences, including cell death or the unrestrained growth that is cancer.

Each part of the cell cycle features its own unique checkpoints. For example, during G1, the cell passes through a critical checkpoint that ensures environmental conditions (including signals from other cells) are favorable for replication. If conditions are not favorable, the cell may enter a resting state known as G0. Some cells remain in G0 for the entire lifetime of the organism in which they reside. For instance, the neurons and skeletal muscle cells of mammals are typically in G0.

Another important checkpoint takes place later in the cell cycle, just before a cell moves from G2 to mitosis. Here, a number of proteins scrutinize the cell's DNA, making sure it is structurally intact and properly replicated. The cell may pause at this point to allow time for DNA repair, if necessary.

Yet another critical cell cycle checkpoint takes place mid-mitosis. This check determines whether the chromosomes in the cell have properly attached to the spindle, or the network of microtubules that will separate them during cell division. This step decreases the possibility that the resulting daughter cells will have unbalanced numbers of chromosomes — a condition called aneuploidy.

How Do Scientists Study the Cell Cycle?

The cell cycle and its system of checkpoint controls show strong evolutionary conservation. As a result, all eukaryotes — from single-celled yeast to complex multicellular vertebrates — pass through the same four phases and same key checkpoints. This universality of the cell cycle and its checkpoint controls allows scientists to use relatively simple model organisms to learn more about cell division in eukaryotes of all types — including humans. In fact, two of the three scientists who received Nobel Prizes for cell cycle research used yeast as the subject of their investigations.

Conclusion

The eukaryotic cell cycle includes four phases necessary for proper growth and division. As a cell moves through each phase, it also passes through several checkpoints. These checkpoints ensure that mitosis occurs only when environmental conditions are favorable and the cellular genome

Concepts of Cell and Molecular Biology

has been precisely replicated. Collectively, this set of checks on division helps prevent chromosomal imbalance in newly produced daughter cells.

Cyclin-Dependent Kinases Regulate Progression through the Cell Cycle

Multiple checkpoints in the eukaryotic cell cycle ensure that division occurs only after sufficient growth and faithful DNA replication, and only when favorable conditions exist. At each checkpoint, numerous proteins engage in a series of carefully coordinated biochemical reactions. This complexity allows for precise regulation of all steps in the cell cycle — and it is essential to preventing the devastating co

What Are Cyclin-Dependent Kinases?

Of the many proteins involved in cell cycle control, cyclin-dependent kinases (CDKs) are among the most important. CDKs are a family of multifunctional enzymes that can modify various protein substrates involved in cell cycle progression. Specifically, CDKs phosphorylate their substrates by transferring phosphate groups from ATP to specific stretches of amino acids in the substrates. Different types of eukaryotic cells contain different types and numbers of CDKs. For example, yeast have only a single CDK, whereas vertebrates have four different ones.

As their name suggests, CDKs require the presence of cyclins to become active. Cyclins are a family of proteins that have no enzymatic activity of their own but activate CDKs by binding to them. CDKs must also be in a particular phosphorylation state — with some sites phosphorylated and others dephosphorylated — in order for activation to occur. Correct phosphorylation depends on the action of other kinases and a second class of enzymes called phosphatases that are responsible for removing phosphate groups from proteins.

How Do CDKs Control the Cell Cycle?

All eukaryotes have multiple cyclins, each of which acts during a specific stage of the cell cycle. (In organisms with multiple CDKs, each CDK is paired with a specific cyclin.) All cyclins are named according to the stage at which they assemble with CDKs. Common classes of cyclins include G1-phase cyclins, G1/S-phase cyclins, S-phase cyclins, and M-phase cyclins. M-phase cyclins form M-CDK complexes and drive the cell's entry into mitosis; G1 cyclins form G1-CDK complexes and guide the cell's progress through the G1phase; and so on.

All CDKs exist in similar amounts throughout the entire cell cycle. In contrast, cyclin manufacture and breakdown varies by stage — with cell cycle progression dependent on the synthesis of new cyclin molecules. Accordingly, cells synthesize G1- and G1/S-cyclins at different times during the G1 phase, and they produce M-cyclin molecules during the G2 phase (Figure 2). Cyclin degradation is equally important for progression through the cell cycle. Specific enzymes break down cyclins at defined times in the cell cycle. When cyclin levels decrease, the corresponding CDKs become inactive. Cell cycle arrest can occur if cyclins fail to degrade.

Which Proteins Do CDKs Modify?

Each of the cyclin-CDK complexes in a cell modifies a specific group of protein substrates. Proper phosphorylation of these substrates must occur at particular times in order for the cell cycle to continue. Because cyclin-CDK complexes recognize multiple substrates, they are able to coordinate the multiple events that occur during each phase of the cell cycle. For example, at the beginning of S phase, S-CDK catalyzes phosphorylation of the proteins that initiate DNA replication by allowing DNA replication complexes to form. Later, during mitosis, M-CDKs phosphorylate a wide range of proteins. These include condensin proteins, which are essential for the extensive condensation of mitotic chromosomes, and lamin proteins, which form a stabilizing network under the nuclear membrane that dissembles during mitosis. M-CDKs also influence the assembly of the mitotic spindle by phosphorylating proteins that regulate microtubule behavior. The net effect of these coordinated phosphorylation reactions is the accurate separation of chromosomes during mitosis.

Conclusion

The llife cycle of a cell is a carefully regulated series of events orchestrated by a suite of enzymes and other proteins. The main regulatory components of cell cycle control are cyclins and CDKs. Depending on the presence and action of these proteins, the cell cycle can be speedy or slow, and it may even halt altogether.

Mitosis Produces Two Daughter Cells with the Same Genetic Makeup

Mitosis is the process in which a eukaryotic cell nucleus splits in two, followed by division of the parent cell into two daughter cells. The word "mitosis" means "threads," and it refers to the threadlike appearance of chromosomes as the cell prepares to divide. Early microscopists were the first to observe these structures, and they also noted the appearance of a specialized network of microtubules during mitosis. These tubules, collectively known as the spindle, extend from structures called centrosomes — with one centrosome located at each of the opposite ends, or poles, of a cell. As mitosis progresses, the microtubules attach to the chromosomes, which have already duplicated their DNA and aligned across the center of the cell. The spindle tubules then shorten and move toward the poles of the cell. As they move, they pull the one copy of each chromosome with them to opposite poles of the cell. This process ensures that each daughter cell will contain one exact copy of the parent cell DNA.

What Are the Phases of Mitosis?

Mitosis consists of five morphologically distinct phases: prophase, prometaphase, metaphase, anaphase, and telophase. Each phase involves characteristic steps in the process of chromosome alignment and separation. Once mitosis is complete, the entire cell divides in two by way of the process called cytokinesis.

What Happens during Prophase?

Prophase is the first stage in mitosis, occurring after the conclusion of the G2 portion of interphase. During prophase, the parent cell chromosomes — which were duplicated during S phase — condense and become thousands of times more compact than they were during interphase. Because each duplicated chromosome consists of two identical sister chromatids joined at a point called the centromere, these structures now appear as X-shaped bodies when viewed under a microscope. Several DNA binding proteins catalyze the condensation process, including cohesin and condensin. Cohesin forms rings that hold the sister chromatids together, whereas condensin forms rings that coil the chromosomes into highly compact forms.

The mitotic spindle also begins to develop during prophase. As the cell's two centrosomes move toward opposite poles, microtubules gradually assemble between them, forming the network that will later pull the duplicated chromosomes apart.

What Happens during Prometaphase?

When prophase is complete, the cell enters prometaphase — the second stage of mitosis. During prometaphase, phosphorylation of nuclear lamins by M-CDK causes the nuclear membrane to break down into numerous small vesicles. As a result, the spindle microtubules now have direct access to the genetic material of the cell.

Each microtubule is highly dynamic, growing outward from the centrosome and collapsing backward as it tries to locate a chromosome. Eventually, the microtubules find their targets and connect to each chromosome at its kinetochore, a complex of proteins positioned at the centromere. The actual number of microtubules that attach to a kinetochore varies between species, but at least one microtubule from each pole attaches to the kinetochore of each chromosome. A tug-of-war then ensues as the chromosomes move back and forth toward the two poles.

What Happens during Metaphase and Anaphase?

As prometaphase ends and metaphase begins, the chromosomes align along the cell equator. Every chromosome has at least two microtubules extending from its kinetochore — with at least one microtubule connected to each pole. At this point, the tension within the cell becomes balanced, and the chromosomes no longer move back and forth. In addition, the spindle is now complete, and three groups of spindle microtubules are apparent. Kinetochore microtubules attach the chromosomes to the spindle pole; interpolar microtubules extend from the spindle pole across the equator, almost to the opposite spindle pole; and astral microtubules extend from the spindle pole to the cell membrane.

Metaphase leads to anaphase, during which each chromosome's sister chromatids separate and move to opposite poles of the cell. Enzymatic breakdown of cohesin — which linked the sister chromatids together during prophase — causes this separation to occur. Upon separation, every chromatid becomes an independent chromosome. Meanwhile, changes in microtubule length provide the mechanism for chromosome movement. More specifically, in the first part of anaphase — sometimes called anaphase A — the kinetochore microtubules shorten and draw the chromosomes toward the spindle poles. Then, in the second part of anaphase — sometimes called anaphase B — the astral microtubules that are anchored to the cell membrane pull the poles further apart and the interpolar microtubules slide past each other, exerting additional pull on the chromosomes (Figure 2).

What Happens during Telophase?

During telophase, the chromosomes arrive at the cell poles, the mitotic spindle disassembles, and the vesicles that contain fragments of the original nuclear membrane assemble around the two sets of chromosomes. Phosphatases then dephosphorylate the lamins at each end of the cell. This dephosphorylation results in the formation of a new nuclear membrane around each group of chromosomes.

When Do Cells Actually Divide?

Cytokinesis is the physical process that finally splits the parent cell into two identical daughter cells. During cytokinesis, the cell membrane pinches in at the cell equator, forming a cleft called the cleavage furrow. The position

Concepts of Cell and Molecular Biology

of the furrow depends on the position of the astral and interpolar microtubules during anaphase.

The cleavage furrow forms because of the action of a contractile ring of overlapping actin and myosin filaments. As the actin and myosin filaments move past each other, the contractile ring becomes smaller, akin to pulling a drawstring at the top of a purse. When the ring reaches its smallest point, the cleavage furrow completely bisects the cell at its center, resulting in two separate daughter cells of equal size (Figure 3).

Conclusion

Mitosis is the process of nuclear division, which occurs just prior to cell division, or cytokinesis. During this multistep process, cell chromosomes condense and the spindle assembles. The duplicated chromosomes then attach to the spindle, align at the cell equator, and move apart as the spindle microtubules retreat toward opposite poles of the cell. Each set of chromosomes is then surrounded by a nuclear membrane, and the parent cell splits into two complete daughter cells.

Tissues Are Organized Communities of Different Cell Types

Within multicellular organisms, tissues are organized communities of cells that work together to carry out a specific function. The exact role of a tissue in an organism depends on what types of cells it contains. For example, the endothelial tissue that lines the human gastrointestinal tract consists of several cell types. Some of these cells absorb nutrients from the digestive contents, whereas others (called goblet cells) secrete a lubricating mucus that helps the contents travel smoothly.

However, the multiple cell types within a tissue don't just have different functions. They also have different transcriptional programs and may well divide at different rates. Proper regulation of these rates is essential to tissue maintenance and repair. The spatial organization of the cells that form a tissue is also central to the tissue's function and survival. This organization depends in part on polarity, or the orientation of particular cells in their place. Of course, external signals from neighboring cells or from the extracellular matrix are also important influences on the arrangement of cells in a tissue.

What Is the Source of New Cells for Tissues?

Without cell division, long-term tissue survival would be impossible. Inside every tissue, cells are constantly replenishing themselves through the process of division, although the rate of turnover may vary widely between different cell types in the same tissue. For example, in adult mammal brains, neurons rarely divide. However, glial cells in the brain continue to divide throughout a mammal's adult life. Mammalian epithelial cells also turn over regularly, typically every few days.

Neurons are not the only cells that lose their ability to divide as they mature. In fact, many differentiated cells lose this ability. To help counteract this loss, tissues maintain stem cells to serve as a reservoir of undifferentiated cells. Stem cells typically have the capacity to mature into many different cell types. Transcription factors — proteins that regulate which genes are transcribed in a cell — appear to be essential to determining the pathway particular stem cells take as they differentiate. For example, both intestinal absorptive cells and goblet cells arise from the same stem cell population, but divergent transcriptional programs cause them to mature into dramatically different cells (Figure 1).

Whenever stem cells are called upon to generate a particular type of cell, they undergo an asymmetric cell division. With asymmetric division, each of the two resulting daughter cells has its own unique life course. In this case, one of the daughter cells has a finite capacity for cell division and begins to differentiate, whereas the other daughter cell remains a stem cell with unlimited proliferative ability.

How Do Non-Growing Tissues Maintain Themselves?

Although most of the tissues in adult organisms maintain a constant size, the cells that make up these tissues are constantly turning over. Therefore, in order for a particular tissue to stay the same size, its rates of cell death and cell division must remain in balance.

A variety of factors can trigger cell death in a tissue. For example, the process of apoptosis, or programmed cell death, selectively removes damaged cells — including those with DNA damage or defective mitochondria. During apoptosis, cellular proteases and nucleases are activated, and cells self-destruct. Cells also monitor the survival factors and negative signals they receive from other cells before initiating programmed cell death. Once apoptosis begins, it proceeds quickly, leaving behind small fragments with recognizable bits of the nuclear material. Specialized cells then rapidly ingest and degrade these fragments, making evidence of apoptosis difficult to detect.

What Cellular Components Support Tissue Structure?

Tissue function depends on more than cell type and proper rates of death and division: It is also a function of cellular arrangement. Both cell junctions and cytoskeletal networks help stabilize tissue architecture. For instance, the cells that make up human epithelial tissue attach to one another through several types of adhesive junctions. Characteristic transmembrane proteins

Concepts of Cell and Molecular Biology

provide the basis for each of the different types of junctions. At these junctions, transmembrane proteins on one cell interact with similar transmembrane proteins on adjacent cells. Special adaptor proteins then connect the resulting assembly to the cytoskeleton of each cell. The many connections formed between junctions and cytoskeletal proteins effectively produces a network that extends over many cells, providing mechanical strength to the epithelium.

The gut endothelium — actually an epithelium that lines the inner surface of the digestive tract — is an excellent example of these structures at work. Here, tight junctions between cells form a seal that prevents even small molecules and ions from moving across the endothelium. As a result, the endothelial cells themselves are responsible for determining which molecules pass from the gut lumen into the surrounding tissues. Meanwhile, adherens junctions based on transmembrane cadherin proteins provide mechanical support to the endothelium. These junctions are reinforced by attachment to an extensive array of actin filaments that underlie the apical — or lumen-facing — membrane. These organized collections of actin filaments also extend into the microvilli, which are the tiny fingerlike projections that protrude from the apical membrane into the gut lumen and increase the surface area available for nutrient absorption. Additional mechanical support comes from desmosomes, which appear as plaque-like structures under the cell membrane, attached to intermediate filaments. In fact, desmosome-intermediate filament networks extend across multiple cells, giving the endothelium sheetlike properties. In addition, within the gut there are stem cells that guarantee a steady supply of new cells that contribute to the multiple cell types necessary for this complex structure to function properly (Figure 2).

How Does the Extracellular Matrix Support Tissue Structure?

The extracellular matrix (ECM) is also critical to tissue structure, because it provides attachment sites for cells and relays information about the spatial position of a cell. The ECM consists of a mixture of proteins and polysaccharides produced by the endoplasmic reticula and Golgi apparatuses of nearby cells. Once synthesized, these molecules move to the appropriate side of the cell — such as the basal or apical face — where they are secreted. Final organization of the ECM then takes place outside the cell.

To understand how the ECM works, consider the two very different sides of the gut endothelium. One side of this tissue faces the lumen, where it comes in contact with digested food. The other side attaches to a specialized ECM support structure called the basal lamina. The basal lamina is composed of collagen and laminin proteins, as well as various other macromolecules.

On this side of the endothelium, adhesive junctions attach cells to the ECM. Transmembrane integrin proteins in the junctions bind components of the ECM and recruit signaling proteins to their cytoplasmic sides. From there, the signals travel to the nucleus of each cell.

Conclusion

Tissues are communities of cells that have functions beyond what any single cell type could accomplish. Healthy tissues require the proper mix of cells, and the cells within them must be oriented correctly and dividing at an appropriate rate. In order to coordinate their function, organization, and rates of death and division, the cells in a tissue are constantly processing and responding to signals from one another and from the ECM around them.

Normal Controls on Cell Division are Lost during Cancer

Cancer cells are cells gone wrong — in other words, they no longer respond to many of the signals that control cellular growth and death. Cancer cells originate within tissues and, as they grow and divide, they diverge ever further from normalcy. Over time, these cells become increasingly resistant to the controls that maintain normal tissue — and as a result, they divide more rapidly than their progenitors and become less dependent on signals from other cells. Cancer cells even evade programmed cell death, despite the fact that their multiple abnormalities would normally make them prime targets for apoptosis. In the late stages of cancer, cells break through normal tissue boundaries and metastasize (spread) to new sites in the body.

How Do Cancer Cells Differ from Normal Cells?

In normal cells, hundreds of genes intricately control the process of cell division. Normal growth requires a balance between the activity of those genes that promote cell proliferation and those that suppress it. It also relies on the activities of genes that signal when damaged cells should undergo apoptosis.

Cells become cancerous after mutations accumulate in the various genes that control cell proliferation. According to research findings from the Cancer Genome Project, most cancer cells possess 60 or more mutations. The challenge for medical researchers is to identify which of these mutations are responsible for particular kinds of cancer. This process is akin to searching for the proverbial needle in a haystack, because many of the mutations present in these cells have little to nothing to do with cancer growth.

Different kinds of cancers have different mutational signatures. However, scientific comparison of multiple tumor types has revealed that certain genes are mutated in cancer cells more often than others. For instance, growth-promoting genes, such as the gene for the signaling protein Ras, are among those most commonly mutated in cancer cells, becoming super-active and producing cells that are too strongly stimulated by growth receptors. Some chemotherapy drugs work to counteract these mutations by blocking the action of growth-signaling proteins. The breast cancer drug Herceptin, for example, blocks overactive receptor tyrosine kinases (RTKs), and the drug Gleevec blocks a mutant signaling kinase associated with chronic myelogenous leukemia.

Other cancer-related mutations inactivate the genes that suppress cell proliferation or those that signal the need for apoptosis. These genes, known as tumor suppressor genes, normally function like brakes on proliferation, and both copies within a cell must be mutated in order for uncontrolled division to occur. For example, many cancer cells carry two mutant copies of the gene that codes for p53, a multifunctional protein that normally senses DNA damage and acts as a transcription factor for checkpoint control genes.

How Do Cancerous Changes Arise?

Gene mutations accumulate over time as a result of independent events. Consequently, the path to cancer involves multiple steps. In fact, many scientists view the progression of cancer as a microevolutionary process.

To understand what this means, consider the following: When a mutation gives a cancer cell a growth advantage, it can make more copies of itself than a normal cell can — and its offspring can outperform their noncancerous counterparts in the competition for resources. Later, a second mutation might provide the cancer cell with yet another reproductive advantage, which in turn intensifies its competitive advantage even more. And, if key checkpoints are missed or repair genes are damaged, then the rate of damage accumulation increases still further. This process continues with every new mutation that offers such benefits, and it is a driving force in the evolution of living things — not just cancer cells (Figure 1, Figure 2).

How Do Cancer Cells Spread to Other Tissues?

During the early stages of cancer, tumors are typically benign and remain confined within the normal boundaries of a tissue. As tumors grow and become malignant, however, they gain the ability to break through these boundaries and invade adjoining tissues.

Invasive cancer cells often secrete proteases that enable them to degrade the extracellular matrix at a tissue's boundary. Proteases also give cancer cells the ability to create new passageways in tissues. For example, they can break down the junctions that join cells together, thereby gaining access to new territories.

Metastasis — literally meaning "new place" — is one of the terminal stages of cancer. In this stage, cancerous cells enter the bloodstream or the lymphatic system and travel to a new location in the body, where they begin to divide and lay the foundation for secondary tumors. Not all cancer cells can metastasize. In order to spread in this way, the cells must have the ability to penetrate the normal barriers of the body so that they can both enter and exit the blood or lymph vessels. Even traveling metastatic cancer cells face challenges when trying to grow in new areas (Figure 3).

Conclusion

Cancer is unchecked cell growth. Mutations in genes can cause cancer by accelerating cell division rates or inhibiting normal controls on the system, such as cell cycle arrest or programmed cell death. As a mass of cancerous cells grows, it can develop into a tumor. Cancer cells can also invade neighboring tissues and sometimes even break off and travel to other parts of the body, leading to the formation of new tumors at those sites.

Normal Controls on Cell Division are Lost during Cancer

Cancer cells are cells gone wrong — in other words, they no longer respond to many of the signals that control cellular growth and death. Cancer cells originate within tissues and, as they grow and divide, they diverge ever further from normalcy. Over time, these cells become increasingly resistant to the controls that maintain normal tissue — and as a result, they divide more rapidly than their progenitors and become less dependent on signals from other cells. Cancer cells even evade programmed cell death, despite the fact that their multiple abnormalities would normally make them prime targets for apoptosis. In the late stages of cancer, cells break through normal tissue boundaries and metastasize (spread) to new sites in the body.

How Do Cancer Cells Differ from Normal Cells?

In normal cells, hundreds of genes intricately control the process of cell division. Normal growth requires a balance between the activity of those genes that promote cell proliferation and those that suppress it. It also relies on the activities of genes that signal when damaged cells should undergo apoptosis.

Cells become cancerous after mutations accumulate in the various genes that control cell proliferation. According to research findings from the Cancer Genome Project, most cancer cells possess 60 or more mutations. The challenge for medical researchers is to identify which of these mutations are responsible for particular kinds of cancer. This process is akin to searching for the proverbial needle in a haystack, because many of the mutations present in these cells have little to nothing to do with cancer growth.

Different kinds of cancers have different mutational signatures. However, scientific comparison of multiple tumor types has revealed that certain genes are mutated in cancer cells more often than others. For instance, growth-promoting genes, such as the gene for the signaling protein Ras, are among those most commonly mutated in cancer cells, becoming super-active and producing cells that are too strongly stimulated by growth receptors. Some chemotherapy drugs work to counteract these mutations by blocking the action of growth-signaling proteins. The breast cancer drug Herceptin, for example, blocks overactive receptor tyrosine kinases (RTKs), and the drug Gleevec blocks a mutant signaling kinase associated with chronic myelogenous leukemia.

Other cancer-related mutations inactivate the genes that suppress cell proliferation or those that signal the need for apoptosis. These genes, known as tumor suppressor genes, normally function like brakes on proliferation, and both copies within a cell must be mutated in order for uncontrolled division to occur. For example, many cancer cells carry two mutant copies of the gene that codes for p53, a multifunctional protein that normally senses DNA damage and acts as a transcription factor for checkpoint control genes.

How Do Cancerous Changes Arise?

Gene mutations accumulate over time as a result of independent events. Consequently, the path to cancer involves multiple steps. In fact, many scientists view the progression of cancer as a microevolutionary process.

To understand what this means, consider the following: When a mutation gives a cancer cell a growth advantage, it can make more copies of itself than a normal cell can — and its offspring can outperform their noncancerous counterparts in the competition for resources. Later, a second mutation might provide the cancer cell with yet another reproductive advantage, which in turn intensifies its competitive advantage even more. And, if key checkpoints are missed or repair genes are damaged, then the rate of damage accumulation increases still further. This process continues with every new mutation that offers such benefits, and it is a driving force in the evolution of living things — not just cancer cells (Figure 1, Figure 2).

How Do Cancer Cells Spread to Other Tissues?

During the early stages of cancer, tumors are typically benign and remain confined within the normal boundaries of a tissue. As tumors grow and become malignant, however, they gain the ability to break through these boundaries and invade adjoining tissues.

Invasive cancer cells often secrete proteases that enable them to degrade the extracellular matrix at a tissue's boundary. Proteases also give cancer cells the ability to create new passageways in tissues. For example, they can break down the junctions that join cells together, thereby gaining access to new territories.

Metastasis — literally meaning "new place" — is one of the terminal stages of cancer. In this stage, cancerous cells enter the bloodstream or the lymphatic system and travel to a new location in the body, where they begin to divide and lay the foundation for secondary tumors. Not all cancer cells can metastasize. In order to spread in this way, the cells must have the ability to penetrate the normal barriers of the body so that they can both enter and exit the blood or lymph vessels. Even traveling metastatic cancer cells face challenges when trying to grow in new areas (Figure 3).

Conclusion

Cancer is unchecked cell growth. Mutations in genes can cause cancer by accelerating cell division rates or inhibiting normal controls on the system, such as cell cycle arrest or programmed cell death. As a mass of cancerous cells grows, it can develop into a tumor. Cancer cells can also invade neighboring tissues and sometimes even break off and travel to other parts of the body, leading to the formation of new tumors at those sites.

TECHNIQUES OF MOLECULAR BIOLOGY

Molecular cloning

One of the most basic techniques of molecular biology to study protein function is molecular cloning. In this technique, DNA coding for a protein of interest is cloned using PCR and/or restriction enzymes into a plasmid (expression vector). A vector has 3 distinctive features: an origin of replication, a multiple cloning site (MCS), and a selective marker usually antibiotic resistance. The origin of replication will have promoter regions upstream from the replication/transcription start site. This plasmid can be inserted into either bacterial or animal cells. Introducing DNA into bacterial cells can be done by transformation via uptake of naked DNA, conjugation via cell-cell contact or by transduction via viral vector. Introducing DNA into eukaryotic cells, such as animal cells, by physical or chemical means is called transfection. Several different transfection techniques are available, such as calcium phosphate transfection, electroporation, microinjection and liposome transfection. The plasmid may be integrated into the genome, resulting in a stable transfection, or may remain independent of the genome, called transient transfection.

DNA coding for a protein of interest is now inside a cell, and the protein can now be expressed. A variety of systems, such as inducible promoters and specific cell-signaling factors, are available to help express the protein of interest at high levels. Large quantities of a protein can then be extracted from the bacterial or eukaryotic cell. The protein can be tested for enzymatic activity under a variety of situations, the protein may be crystallized so its tertiary structure can be studied, or, in the pharmaceutical industry, the activity of new drugs against the protein can be studied.[citation needed]

Polymerase chain reaction (PCR

Polymerase chain reaction is an extremely versatile technique for copying DNA. In brief, PCR allows a specific DNA sequence to be copied or modified in predetermined ways. The reaction is extremely powerful and under perfect conditions could amplify 1 DNA molecule to become 1.07 billion molecules in less than 2 hours. The PCR technique can be used to introduce restriction enzyme sites to ends of DNA molecules, or to mutate particular bases of DNA, the latter is a method referred to as site-directed mutagenesis. PCR can also be used to determine whether a particular DNA fragment is found in a cDNA library. PCR has many variations, like reverse transcription PCR (RT-PCR) for amplification of RNA, and, more recently, quantitative PCR which allow for quantitative measurement of DNA or RNA molecules.[8][9]

Gel electrophoresis

Gel electrophoresis is one of the principal tools of molecular biology. The basic principle is that DNA, RNA, and proteins can all be separated by means of an electric field and size. In agarose gel electrophoresis, DNA and RNA can be separated on the basis of size by running the DNA through an electrically charged agarose gel. Proteins can be separated on the basis of size by using an SDS-PAGE gel, or on the basis of size and their electric charge by using what is known as a 2D gel electrophoresis.

Macromolecule blotting and probing

The terms northern, western and eastern blotting are derived from what initially was a molecular biology joke that played on the term Southern blotting, after the technique described by Edwin Southern for the hybridisation of blotted DNA. Patricia Thomas, developer of the RNA blot which then became known as the northern blot, actually didn't use the term.

Southern blotting

Named after its inventor, biologist Edwin Southern, the Southern blot is a method for probing for the presence of a specific DNA sequence within a DNA sample. DNA samples before or after restriction enzyme (restriction endonuclease) digestion are separated by gel electrophoresis and then transferred to a membrane by blotting via capillary action. The membrane is then exposed to a labeled DNA probe that has a complement base sequence to the sequence on the DNA of interest. Southern blotting is less commonly used in laboratory science due to the capacity of other techniques, such as PCR, to detect specific DNA sequences from DNA samples. These blots are still used for some applications, however, such as measuring transgene copy number in transgenic mice or in the engineering of gene knockout embryonic stem cell lines.

Northern blotting

The northern blot is used to study the expression patterns of a specific type of RNA molecule as relative comparison among a set of different samples of RNA. It is essentially a combination of denaturing RNA gel electrophoresis, and a blot. In this process RNA is separated based on size and is then transferred to a membrane that is then probed with a labeled complement of a sequence of interest. The results may be visualized through a variety of ways depending on the label used; however, most result in the revelation of bands representing the sizes of the RNA detected in sample. The intensity of these bands is related to the amount of the target RNA in the samples analyzed. The procedure is commonly used to study when and how much gene expression is occurring by measuring how much of that RNA is present in different samples. It is one of the most basic tools for determining at what time, and under what conditions, certain genes are expressed in living tissues.

Western blotting

In western blotting, proteins are first separated by size, in a thin gel sandwiched between two glass plates in a technique known as SDS-PAGE. The proteins in the gel are then transferred to a polyvinylidene fluoride (PVDF), nitrocellulose, nylon, or other support membrane. This membrane can then be probed with solutions of antibodies. Antibodies that specifically bind to the protein of interest can then be visualized by a variety of techniques, including colored products, chemiluminescence, or autoradiography. Often, the antibodies are labeled with enzymes. When a chemiluminescent substrate is exposed to the enzyme it allows detection. Using western blotting techniques allows not only detection but also quantitative analysis. Analogous methods to western blotting can be used to directly stain specific proteins in live cells or tissue sections.

Eastern blotting

The Eastern blotting technique is used to detect post-translational modification of proteins. Proteins blotted on to the PVDF or nitrocellulose membrane are probed for modifications using specific substrates.

Microarrays

DNA microarray is a collection of spots attached to a solid support such as a microscope slide where each spot contains one or more single-stranded DNA oligonucleotide fragment. Arrays make it possible to put down large quantities of very small (100 micrometre diameter) spots on a single slide. Each spot has a DNA fragment molecule that is complementary to a single DNA sequence. A variation of this technique allows the gene expression of an organism at a particular stage in development to be qualified (expression profiling). In this technique the RNA in a tissue is isolated and converted to labeled cDNA. This cDNA is then hybridized to the fragments on the array and visualization of the hybridization can be done. Since multiple arrays can be made with exactly the same position of fragments they are particularly useful for comparing the gene expression of two different tissues, such as a healthy and cancerous tissue. Also, one can measure what genes are expressed and how that expression changes with time or with other factors. There are many different ways to fabricate microarrays; the most common are silicon chips, microscope slides with spots of ~100 micrometre diameter, custom arrays, and arrays with larger spots on porous membranes (macroarrays). There can be anywhere from 100 spots to more than 10,000 on a given array. Arrays can also be made with molecules other than DNA.

Allele-specific oligonucleotide

Allele-specific oligonucleotide (ASO) is a technique that allows detection of single base mutations without the need for PCR or gel electrophoresis. Short (20-25 nucleotides in length), labeled probes are exposed to the non-fragmented target DNA, hybridization occurs with high specificity due to the short length of the probes and even a single base change will hinder hybridization. The target DNA is then washed and the labeled probes that didn't hybridize are removed. The target DNA is then analyzed for the presence of the probe via radioactivity or fluorescence. In this experiment, as in most molecular biology techniques, a control must be used to ensure successful experimentation.

In molecular biology, procedures and technologies are continually being developed and older technologies abandoned. For example, before the advent of DNA gel electrophoresis (agarose or polyacrylamide), the size of DNA molecules was typically determined by rate sedimentation in sucrose gradients, a slow and labor-intensive technique requiring expensive instrumentation; prior to sucrose gradients, viscometry was used. Aside from their historical interest, it is often worth knowing about older technology, as it is occasionally useful to solve another new problem for which the newer technique is inappropriate.

Conclusion

Significant information on the chemical structure, macromolecular characteristics, biological properties and medical application of hyaluronan has accumulated in international literature. The understanding of the molecular mechanisms of hyaluronic acid in the body is still limited but it is clear that it has played a crucial role in the historical development (phylogenicity) of chordates and ontogenicity of modern higher vertebrates. It is part of the main different types of intercellular substance of connective tissue, is present in large amounts in the vitreous body of the eye, synovial fluid of joints, oviducts, Wharton's jelly of the umbilical cord, skin, walls of arteries and veins, heart valves, the glomerular basement membrane of the kidneys and so on. Since the discovery of hyaluronan, there has been a significant evolution of the views on this biopolymer. The polysaccharide was initially considered to be a passive structural component of the extracellular matrix. Now, however, it has become apparent that it is dynamically involved into many biological processes, from reproduction modulation, cell migration and differentiation during embryogenesis, to regulation of processes of inflammation and wound healing and cancer cell metastasis. Hyaluronan has multiple physiological functions in the body. It is the basis of the functioning of the mucolytic system (hyaluronan – hyaluronidase) that determines, in particular, permeability of tissue and vascular circulatory system, the resistance to penetration of infection and substance filtration in the kidneys. Hyaluronan acts as a structure that forms compounds and stores water in the gel-like matrix of the extracellular matrix. This function provides mechanical support, tissue turgor, resistance to compressive pressure and determines the damping and anti-friction properties of the synovial fluid of joints. Hyaluronan is involved in the fertilization of oocytes, as well as their division, migration, maintenance of the cells in the differentiation state and transitions of the differentiated cells into the cell cycle and back. Hyaluronan participates in the functioning of cell signalling systems.

Such a wide variety of the biological properties of hyaluronan are first of all functions of molecular weight, polymorphism of the different structural forms, physicochemical properties of molecules of different molecular weight and the ionic environment and concentration of the biopolymer in tissues and organs. Upon cleavage of polysaccharide macromolecules, the different polymorphous structures with different structures and functional properties are formed. The formed fragments have broad and often contradictory functions and properties. Short biopolymer chains often behave as 'alarm signals' whereas longer fragments act as triggers for the signalling systems. For example, tetrasaccharides induce heat shock proteins that are supposed to heal the cell damage and inhibit apoptosis, leading to cell death. Similar biological properties of hyaluronan degradation products, which help cell preservation, represent the brilliant example of extending Le Chatelier's principle on biological systems. Such properties confirm the validity of the concept that the degradation products of biopolymers promote restoration and/or preservation of system homeostasis and can be used as control factors of the regenerative process in the environment in which they are present. Hence, this is a natural way, by which a strategy for medical use of known products based on hyaluronan could be developed and the new generation of the therapeutic products could be created. In this book we have tried to consider many of issues with different degrees of complexity and completeness of the presented material. It seems that by choosing one particular problem, for example, the biological role of hyaluronan, we could protect the reader from the inexhaustible variety of other no less interesting but nevertheless different areas of knowledge and data. However, we intentionally used a multidisciplinary approach while being aware that some issues are described superficially and require address to additional special literature.

5

Gene Expression and Regulation
(Recombinant of DNA and RNA Technology)

INTRODUCTION

Gene expression is the process by which the genetic code - the nucleotide sequence - of a gene is used to direct protein synthesis and produce the structures of the cell. Genes that code for amino acid sequences are known as 'structural genes'.

The process of gene expression involves two main stages:

Transcription: the production of messenger RNA (mRNA) by the enzyme RNA polymerase, and the processing of the resulting mRNA molecule.

Translation: the use of mRNA to direct protein synthesis, and the subsequent post-translational processing of the protein molecule.

Some genes are responsible for the production of other forms of RNA that play a role in translation, including transfer RNA (tRNA) and ribosomal RNA (rRNA).

- Exons. Exons code for amino acids and collectively determine the amino acid sequence of the protein product. It is these portions of the gene that are represented in final mature mRNA molecule.
- Introns. Introns are portions of the gene that do not code for amino acids, and are removed (spliced) from the mRNA molecule before translation

But according to the Wikipedia: Regulation of gene expression includes a wide range of mechanisms that are used by cells to increase or decrease the production of specific gene products (protein or RNA), and is informally termed gene regulation. Sophisticated programs of gene expression are widely observed in biology, for example to trigger developmental pathways, respond to environmental stimuli, or adapt to new food sources. Virtually any step of gene expression can be modulated, from transcriptional initiation, to RNA processing, and to the post-translational modification of a protein. Often, one gene regulator controls another, and so on, in a gene regulatory network.

Gene regulation is essential for viruses, prokaryotes and eukaryotes as it increases the versatility and adaptability of an organism by allowing the cell to express protein when needed. Although as early as 1951, Barbara McClintock showed interaction between two genetic loci, Activator (Ac) and Dissociator (Ds), in the color formation of maize seeds, the first discovery of a gene regulation system is widely considered to be the identification in 1961 of the lac operon, discovered by Jacques Monod, in which some enzymes involved in lactose metabolism are expressed by E. coli only in the presence of lactose and absence of glucose.

In multicellular organisms, gene regulation drives cellular differentiation and morphogenesis in the embryo, leading to the creation of different cell types that possess different gene expression profiles from the same genome sequence. This explains how evolution actually works at a molecular level, and is central to the science of evolutionary developmental biology ("evo-devo").

The initiating event leading to a change in gene expression includes activation or deactivation of receptors.

Gene Expression and Regulation

How does a gene, which consists of a string of DNA hidden in a cell's nucleus, know when it should express itself? How does this gene cause the production of a string of amino acids called a protein? How do different types of cells know which types of proteins they must manufacture? The answers to such questions lie in the study of gene expression. Thus, this collection or articles begins by showing how a quiet, well-guarded string of DNA is expressed to make RNA, and how the messenger RNA is translated from nucleic acid coding to protein coding to form a protein. Along the way, the article set also examines the nature of the genetic code, how the elements of code were predicted, and how the actual codons were determined.

Next, we turn to the regulation of genes. Genes can't control an organism on their own; rather, they must interact with and respond to the organism's environment. Some genes are constitutive, or always "on," regardless of environmental conditions. Such genes are among the most important elements of a cell's genome, and they control the ability of DNA to replicate, express itself, and repair itself. These genes also control protein synthesis and much of an organism's central metabolism. In contrast, regulated genes are needed only occasionally — but how do these genes get turned "on" and "off"? What specific molecules control when they are expressed?

It turns out that the regulation of such genes differs between prokaryotes and eukaryotes. For prokaryotes, most regulatory proteins are negative and therefore turn genes off. Here, the cells rely on protein–small molecule binding, in which a ligand or small molecule signals the state of the cell and whether gene expression is needed. The repressor or activator protein binds near its regulatory target: the gene. Some regulatory proteins must have a ligand attached to them to be able to bind, whereas others are unable to bind when attached to a ligand. In prokaryotes, most regulatory proteins are specific to one gene, although there are a few proteins that act more widely. For instance, some repressors bind near the start of mRNA production for an entire operon, or cluster of coregulated genes. Furthermore, some repressors have a fine-tuning system known as attenuation, which uses mRNA structure to stop both transcription and translation depending on the concentration of an operon's end-product enzymes. (In eukaryotes, there is no exact equivalent of attenuation, because transcription occurs in the nucleus and translation occurs in the cytoplasm, making this sort of coordinated effect impossible.) Yet another layer of prokaryotic regulation affects the structure of RNA polymerase, which turns on large groups of genes. Here, the sigma factor of RNA polymerase changes several times to produce heat- and desiccation-resistant spores. Here, the articles on prokaryotic regulation delve into each of these topics, leading to primary literature in many cases.

For eukaryotes, cell-cell differences are determined by expression of different sets of genes. For instance, an undifferentiated fertilized egg looks and acts quite different from a skin cell, a neuron, or a muscle cell because of differences in the genes each cell expresses. A cancer cell acts different from a normal cell for the same reason: It expresses different genes. (Using microarray analysis, scientists can use such differences to assist in diagnosis and selection of appropriate cancer treatment.) Interestingly, in eukaryotes, the default state of gene expression is "off" rather than "on," as in prokaryotes. Why is this the case? The secret lies in chromatin, or the complex of DNA and histone proteins found within the cellular nucleus. The histones are among the most evolutionarily conserved proteins known; they are vital for the well-being of eukaryotes and brook little change. When a specific gene is tightly bound with histone, that gene is "off." But how, then, do eukaryotic genes manage to escape this silencing? This is where the histone code comes into play. This code includes modifications of the histones' positively charged amino acids to create some domains in which DNA is more open and others in which it is very tightly bound up. DNA methylation is one mechanism that appears to be coordinated with histone modifications, particularly those that lead to silencing of gene expression. Small noncoding RNAs such as

Gene Expression and Regulation

RNAi can also be involved in the regulatory processes that form "silent" chromatin. On the other hand, when the tails of histone molecules are acetylated at specific locations, these molecules have less interaction with DNA, thereby leaving it more open. The regulation of the opening of such domains is a hot topic in research. For instance, researchers now know that complexes of proteins called chromatin remodeling complexes use ATP to repackage DNA in more open configurations. Scientists have also determined that it is possible for cells to maintain the same histone code and DNA methylation patterns through many cell divisions. This persistence without reliance on base pairing is called epigenetics, and there is abundant evidence that epigenetic changes cause many human diseases.

For transcription to occur, the area around a prospective transcription zone needs to be unwound. This is a complex process requiring the coordination of histone modifications, transcription factor binding and other chromatin remodeling activities. Once the DNA is open, specific DNA sequences are then accessible for specific proteins to bind. Many of these proteins are activators, while others are repressors; in eukaryotes, all such proteins are often called transcription factors (TFs). Each TF has a specific DNA binding domain that recognizes a 6-10 base-pair motif in the DNA, as well as an effector domain. In the test tube, scientists can find a footprint of a TF if that protein binds to its matching motif in a piece of DNA. They can also see whether TF binding slows the migration of DNA in gel electrophoresis.

For an activating TF, the effector domain recruits RNA polymerase II, the eukaryotic mRNA-producing polymerase, to begin transcription of the corresponding gene. Some activating TFs even turn on multiple genes at once. All TFs bind at the promoters just upstream of eukaryotic genes, similar to bacterial regulatory proteins. However, they also bind at regions called enhancers, which can be oriented forward or backwards and located upstream or downstream or even in the introns of a gene, and still activate gene expression. Because many genes are coregulated, studying gene expression across the whole genome via microarrays or massively parallel sequencing allows investigators to see which groups of genes are coregulated during differentiation, cancer, and other states and processes.

Most eukaryotes also make use of small noncoding RNAs to regulate gene expression. For example, the enzyme Dicer finds double-stranded regions of RNA and cuts out short pieces that can serve in a regulatory role. Argonaute is another enzyme that is important in regulation of small noncoding RNA–dependent systems. Here we offfer an introductory article on these RNAs, but more content is needed; please contact the editors if you are interested in contributing.

Imprinting is yet another process involved in eukaryotic gene regulation; this process involves the silencing of one of the two alleles of a gene for a cell's entire life span. Imprinting affects a minority of genes, but several important growth regulators are included. For some genes, the maternal copy is always silenced, while for different genes, the paternal copy is always silenced. The epigenetic marks placed on these genes during egg or sperm formation are faithfully copied into each subsequent cell, thereby affecting these genes throughout the life of the organism.

Still another mechanism that causes some genes to be silenced for an organism's entire lifetime is X inactivation. In female mammals, for instance, one of the two copies of the X chromosome is shut off and compacted greatly. This shutoff process requires transcription, the participation of two noncoding RNAs (one of which coats the inactive X chromosome), and the participation of a DNA-binding protein called CTCF. As the possible role of regulatory noncoding RNAs in this process is investigated, more information regarding X inactivation will no doubt be discovered.

Gene expression can be regulated by various cellular processes with the aim to control the amount and nature of the expressed genes. Expression of genes can be controlled with the help of regulatory proteins at numerous levels. These regulatory proteins bind to DNA and send signals that indirectly control the rate of gene expression. The up-regulation of a gene refers to an increase in expression of a gene whilst down-regulation refers to the decrease in expression of a gene.

Gene expression can be regulated by various cellular processes with the aim to control the amount and nature of the expressed genes. Expression of genes can be controlled with the help of regulatory proteins at numerous levels. These regulatory proteins bind to DNA and send signals that indirectly control the rate of gene expression. The up-regulation of a gene refers to an increase in expression of a gene whilst down-regulation refers to the decrease in expression of a gene.

In other words Gene expression is the process by which the genetic code - the nucleotide sequence - of a gene is used to direct protein synthesis and produce the structures of the cell. Genes that code for amino acid sequences are called as Structural genes. The process of gene expression involves two main stages as Transcription: the production of messenger RNA (mRNA) by the enzyme RNA polymerase, and the processing of the resulting mRNA molecule. Translation: the use of mRNA to direct protein synthesis, and the subsequent post-translational processing of the protein molecule. Any step of gene expression may be modulated, from the DNA-RNA transcription step to post-translational modification of a protein.

Gene Expression and Regulation

The control of gene expression is more complex in eukaryotes than in prokaryotes. This is because of the presence of a nuclear membrane in eukaryotes which separates the genetic material from the translation machinery.

This necessitates some additional steps such as messenger RNA (mRNA) transport and resultant eukaryotic gene regulation at many different points. In contrast, prokaryotes lack a clearly defined nucleus hence the key point at which their gene regulation occurs is during transcriptional initiation.

Actually Gene expression is the process by which information from a gene is used in the synthesis of a functional gene product. These products are usually proteins which functions as enzymes, hormones and receptors. Genes which do not code for proteins such as ribosomal RNA or transfer RNA code for functional RNA products. Gene expression is the process by which the genetic code the nucleotide sequence of a gene is used to direct protein synthesis and produce the structures of the cell. Genes that code for amino acid sequences are called as structural genes.

Regulated stages of gene expression

Any step of gene expression may be modulated, from the DNA-RNA transcription step to post-translational modification of a protein. The following is a list of stages where gene expression is regulated, the most extensively utilised point is Transcription Initiation:

Epigenome

An epigenome consists of a record of the chemical changes to the DNA and histone proteins of an organism; these changes can be passed down to an organism's offspring via transgenerational epigenetic inheritance. Changes to the epigenome can result in changes to the structure of chromatin and changes to the function of the genome.

The epigenome is involved in regulating gene expression, development, tissue differentiation, and suppression of transposable elements. Unlike the underlying genome which is largely static within an individual, the epigenome can be dynamically altered by environmental conditions

The epigenome is a multitude of chemical compounds that can tell the genome what to do. The human genome is the complete assembly of DNA (deoxyribonucleic acid)-about 3 billion base pairs - that makes each individual unique.

What is the epigenome?

DNA modifications that do not change the DNA sequence can affect gene activity. Chemical compounds that are added to single genes can regulate their activity; these modifications are known as epigenetic changes. The epigenome comprises all of the chemical compounds that have been added to the entirety of one's DNA (genome) as a way to regulate the activity (expression) of all the genes within the genome. The chemical compounds of the epigenome are not part of the DNA sequence, but are on or attached to DNA ("epi-" means above in Greek). Epigenomic modifications remain as cells divide and in some cases can be inherited through the generations. Environmental influences, such as a person's diet and exposure to pollutants, can also impact the epigenome.

Epigenetic changes can help determine whether genes are turned on or off and can influence the production of proteins in certain cells, ensuring that only necessary proteins are produced. For example, proteins that promote bone growth are not produced in muscle cells. Patterns of epigenome modification vary among individuals, different tissues within an individual, and even different cells.

A common type of epigenomic modification is called methylation. Methylation involves attaching small molecules called methyl groups, each consisting of one carbon atom and three hydrogen atoms, to segments of DNA. When methyl groups are added to a particular gene, that gene is turned off or silenced, and no protein is produced from that gene.

Because errors in the epigenetic process, such as modifying the wrong gene or failing to add a compound to a gene, can lead to abnormal gene activity or inactivity, they can cause genetic disorders. Conditions including cancers, metabolic disorders, and degenerative disorders have all been found to be related to epigenetic errors.

Scientists continue to explore the relationship between the genome and the chemical compounds that modify it. In particular, they are studying what effect the modifications have on gene function, protein production, and human health.

In other words The epigenome is a multitude of chemical compounds that can tell the genome what to do. The human genome is the complete assembly of DNA (deoxyribonucleic acid)-about 3 billion base pairs - that makes each individual unique. DNA holds the instructions for building the proteins that carry out a variety of functions in a cell. The epigenome is made up of chemical compounds and proteins that can attach to DNA and direct such actions as turning genes on or off, controlling the production of proteins in particular cells.

When epigenomic compounds attach to DNA and modify its function, they are said to have "marked" the genome. These marks do not change the sequence of the DNA. Rather, they change the way cells use the DNA's

Gene Expression and Regulation

instructions. The marks are sometimes passed on from cell to cell as cells divide. They also can be passed down from one generation to the next.

What does the epigenome do?

A human being has trillions of cells, specialized for different functions in muscles, bones and the brain, and each of these cells carries essentially the same genome in its nucleus. The differences among cells are determined by how and when different sets of genes are turned on or off in various kinds of cells. Specialized cells in the eye turn on genes that make proteins that can detect light, while specialized cells in red blood cells make proteins that carry oxygen from the air to the rest of the body. The epigenome controls many of these changes to the genome.

What makes up the epigenome?

The epigenome is the set of chemical modifications to the DNA and DNA-associated proteins in the cell, which alter gene expression, and are heritable (via meiosis and mitosis). The modifications occur as a natural process of development and tissue differentiation, and can be altered in response to environmental exposures or disease.

The first type of mark, called DNA methylation, directly affects the DNA in a genome. In this process, proteins attach chemical tags called methyl groups to the bases of the DNA molecule in specific places. The methyl groups turn genes on or off by affecting interactions between the DNA and other proteins. In this way, cells can remember which genes are on or off.

The second kind of mark, called histone modification, affects DNA indirectly. DNA in cells is wrapped around histone proteins, which form spool-like structures that enable DNA's very long molecules to be wound up neatly into chromosomes inside the cell nucleus. Proteins can attach a variety of chemical tags to histones. Other proteins in cells can detect these tags and determine whether that region of DNA should be used or ignored in that cell.

Is the epigenome inherited?

The genome is passed from parents to their offspring and from cells, when they divide, to their next generation. Much of the epigenome is reset when parents pass their genomes to their offspring; however, under some circumstances, some of the chemical tags on the DNA and histones of eggs and sperm may be passed on to the next generation. When cells divide, often much of the epigenome is passed on to the next generation of cells, helping the cells remain specialized.

What is imprinting?

The human genome contains two copies of every gene-one copy inherited from the mother and one from the father. For a small number of genes, only the copy from the mother gets switched on; for others, only the copy from the father is turned on. This pattern is called imprinting. The epigenome distinguishes between the two copies of an imprinted gene and determines which is switched on.

Some diseases are caused by abnormal imprinting. They include Beckwith-Wiedemann syndrome, a disorder associated with body overgrowth and increased risk of cancer; Prader-Willi syndrome, associated with poor muscle tone and constant hunger, leading to obesity; and Angelman syndrome, which leads to intellectual disability, as well motion difficulties.

Can the epigenome change?

Although all cells in the body contain essentially the same genome, the DNA marked by chemical tags on the DNA and histones gets rearranged when cells become specialized. The epigenome can also change throughout a person's lifetime.

What makes the epigenome change?

Lifestyle and environmental factors (such as smoking, diet and infectious disease) can expose a person to pressures that prompt chemical responses. These responses, in turn, often lead to changes in the epigenome, some of which can be damaging. However, the ability of the epigenome to adjust to the pressures of life appears to be required for normal human health. Some human diseases are caused by malfunctions in the proteins that "read" and "write" epigenomic marks.

How do changes in the epigenome contribute to cancer?

Cancers are caused by changes in the genome, the epigenome, or both. Changes in the epigenome can switch on or off genes involved in cell growth or the immune response. These changes can lead to uncontrolled growth, a hallmark of cancer, or to a failure of the immune system to destroy tumors.

In a type of brain tumor called glioblastoma, doctors have had some success in treating patients with the drug temozolomide, which kills cancer cells by adding methyl groups to DNA. In some cases, methylation has a welcome secondary effect: it blocks a gene that counteracts temozolomide. Glioblastoma patients whose tumors have such methylated genes are far more likely to respond to temozolomide than those with unmethylated genes.

Changes in the epigenome also can activate growth-promoting genes in stomach cancer, colon cancer and the most common type of kidney cancer. In some other cancers, changes in the epigenome silence genes that normally serve to keep cell growth in check.

To compile a complete list of possible epigenomic changes that can lead to cancer, researchers in The Cancer Genome Atlas (TCGA) Network, which is supported by the National Institutes of Health (NIH), are comparing the genomes and epigenomes of normal cells with those of cancer cells. Among other things, they are looking for changes in the DNA sequence and changes in the number of methyl groups on the DNA.

Understanding all the changes that turn a normal cell into a cancer cell will speed efforts to develop new and better ways of diagnosing, treating and preventing cancer.

How are researchers exploring the epigenome?

In a field of study known as epigenomics, researchers are trying to chart the locations and understand the functions of all the chemical tags that mark the genome.

Until recently, scientists thought that human diseases were caused mainly by changes in DNA sequence, infectious agents such as bacteria and viruses, or environmental agents. Now, however, researchers have demonstrated that changes in the epigenome also can cause, or result from, disease. Epigenomics, thus, has become a vital part of efforts to better understand the human body and to improve human health. Epigenomic maps may someday enable doctors to determine an individual's health status and tailor a patient's response to therapies.

As part of the ENCODE (ENCyclopedia Of DNA Elements) project-which aims to catalog the working parts of the genome-the National Human Genome Research Institute is funding researchers to make epigenomic maps of various cell types. Other NIH-supported investigators have developed a number of epigenomic maps from several human organs and tissues. These NIH projects are part of an international effort to understand how epigenomics could lead to better prevention, diagnosis and treatment of disease.

Nutrition & the Epigenome

Unlike behavior or stress, diet is one of the more easily studied, and therefore better understood, environmental factors in epigenetic change.

The nutrients we extract from food enter metabolic pathways where they are manipulated, modified, and molded into molecules the body can use. One such pathway is responsible for making methyl groups - important epigenetic tags that silence genes.

Familiar nutrients like folic acid, B vitamins, and SAM-e (S-Adenosyl methionine, a popular over-the-counter supplement) are key components of this methyl-making pathway. Diets high in these methyl-donating nutrients can rapidly alter gene expression, especially during early development when the epigenome is first being established.

Transcription

Transcription is the first step of gene expression, in which a particular segment of DNA is copied into RNA (especially mRNA) by the enzyme RNA polymerase. Both DNA and RNA are nucleic acids, which use base pairs of nucleotides as a complementary language.

A gene is a stretch of DNA that encodes information. Genomic DNA consists of two antiparallel and reverse complementary strands, each having 5' and 3' ends. With respect to a gene, the two strands may be labeled the "template strand," which serves as a blueprint for the production of an RNA transcript, and the "coding strand," which includes the DNA version of the transcript sequence. (Perhaps surprisingly, the "coding strand" is not physically involved in the coding process because it is the "template strand" that is read during transcription.)

The production of the RNA copy of the DNA is called transcription, and is performed in the nucleus by RNA polymerase, which adds one RNA nucleotide at a time to a growing RNA strand as per the complementarity law of the bases. This RNA is complementary to the template 3' ? 5' DNA strand, which is itself complementary to the coding 5' ? 3' DNA strand. Therefore, the resulting 5' ? 3' RNA strand is identical to the coding DNA strand with the exception that thymines (T) are replaced with uracils (U) in the RNA. A coding DNA strand reading "ATG" is indirectly transcribed through the non-coding strand as "AUG" in RNA.

In prokaryotes, transcription is carried out by a single type of RNA polymerase, which needs a DNA sequence called a Pribnow box as well as a sigma factor (s factor) to start transcription. In eukaryotes, transcription is performed by three types of RNA polymerases, each of which needs a special DNA sequence called the promoter and a set of DNA-binding proteins—transcription factors—to initiate the process. RNA polymerase I is responsible for transcription of ribosomal RNA (rRNA) genes. RNA polymerase II (Pol II) transcribes all protein-coding genes but also some non-coding RNAs (e.g., snRNAs, snoRNAs or long non-coding RNAs). Pol II includes a C-terminal domain (CTD) that is rich in serine residues. When these residues are phosphorylated, the CTD binds to various protein factors that promote transcript maturation and modification. RNA polymerase III transcribes 5S rRNA, transfer RNA (tRNA) genes, and some small non-coding RNAs (e.g.,

Gene Expression and Regulation

7SK). Transcription ends when the polymerase encounters a sequence called the terminator.

In transcription, the DNA sequence of a gene is transcribed (copied out) to make an RNA molecule.

What makes death cap mushrooms deadly? These mushrooms get their lethal effects by producing one specific toxin, which attaches to a crucial enzyme in the human body: RNA polymerase.

RNA polymerase is crucial because it carries out transcription, the process of copying DNA (deoxyribonucleic acid, the genetic material) into RNA (ribonucleic acid, a similar but more short-lived molecule).

Transcription is an essential step in using the information from genes in our DNA to make proteins. Proteins are the key molecules that give cells structure and keep them running. Blocking transcription with mushroom toxin causes liver failure and death, because no new RNAs—and thus, no new proteins—can be made.^22start superscript, 2, end superscript

Transcription is essential to life, and understanding how it works is important to human health. Let's take a closer look at what happens during transcription.

Transcription overview

Transcription is the first step of gene expression. During this process, the DNA sequence of a gene is copied into RNA.

Before transcription can take place, the DNA double helix must unwind near the gene that is getting transcribed. The region of opened-up DNA is called a transcription bubble.

Transcription uses one of the two exposed DNA strands as a template; this strand is called the template strand. The RNA product is complementary to the template strand and is almost identical to the other DNA strand, called the nontemplate (or coding) strand. However, there is one important difference: in the newly made RNA, all of the T nucleotides are replaced with U nucleotides.

The site on the DNA from which the first RNA nucleotide is transcribed is called the +1+1plus, 1 site, or the initiation site. Nucleotides that come before the initiation site are given negative numbers and said to be upstream. Nucleotides that come after the initiation site are marked with positive numbers and said to be downstream.

If the gene that's transcribed encodes a protein (which many genes do), the RNA molecule will be read to make a protein in a process called translation.

RNA polymerase

RNA polymerases are enzymes that transcribe DNA into RNA. Using a DNA template, RNA polymerase builds a new RNA molecule through base pairing. For instance, if there is a G in the DNA template, RNA polymerase will add a C to the new, growing RNA strand.

Stages of transcription

Transcription of a gene takes place in three stages: initiation, elongation, and termination. Here, we will briefly see how these steps happen in bacteria.

Initiation. RNA polymerase binds to a sequence of DNA called the promoter, found near the beginning of a gene. Each gene (or group of co-transcribed genes, in bacteria) has its own promoter. Once bound, RNA polymerase separates the DNA strands, providing the single-stranded template needed for transcription.

Elongation. One strand of DNA, the template strand, acts as a template for RNA polymerase. As it "reads" this template one base at a time, the polymerase builds an RNA molecule out of complementary nucleotides, making a chain that grows from 5' to 3'. The RNA transcript carries the same information as the non-template (coding) strand of DNA, but it contains the base uracil (U) instead of thymine (T).

Termination. Sequences called terminators signal that the RNA transcript is complete. Once they are transcribed, they cause the transcript to be released from the RNA polymerase. An example of a termination mechanism involving formation of a hairpin in the RNA.

Eukaryotic RNA modifications

In bacteria, RNA transcripts can act as messenger RNAs (mRNAs) right away. In eukaryotes, the transcript of a protein-coding gene is called a pre-mRNA and must go through extra processing before it can direct translation.
- Eukaryotic pre-mRNAs must have their ends modified, by addition of a 5' cap (at the beginning) and 3' poly-A tail (at the end).
- Many eukaryotic pre-mRNAs undergo splicing. In this process, parts of the pre-mRNA (called introns) are chopped out, and the remaining pieces (called exons) are stuck back together.

The pre-mRNA still contains both exons and introns. Along the length of the mRNA, there is an alternating pattern of exons and introns: Exon 1 - Intron 1 - Exon 2 - Intron 2 - Exon 3. Each consists of a stretch of RNA nucleotides.

During splicing, the introns are removed from the pre-mRNA, and the exons are stuck together to form a mature mRNA.

Gene Expression and Regulation 165

Bottom of image: Mature mRNA that does not contain the intron sequences (Exon 1 - Exon 2 - Exon 3 only).

End modifications increase the stability of the mRNA, while splicing gives the mRNA its correct sequence. (If the introns are not removed, they'll be translated along with the exons, producing a "gibberish" polypeptide.)

To learn more about pre-mRNA modifications in eukaryotes, check out the article on pre-mRNA processing.

Transcription happens for individual genes

Not all genes are transcribed all the time. Instead, transcription is controlled individually for each gene (or, in bacteria, for small groups of genes that are transcribed together). Cells carefully regulate transcription, transcribing just the genes whose products are needed at a particular moment.

For example, the diagram below shows a "snapshot" of an imaginary cell's RNAs at a given moment in time. In this cell, genes 1, 2 and 3, are transcribed, while gene 4 is not. Also, genes 1, 2, and 3 are transcribed at different levels, meaning that different numbers of RNA molecules are made for each.

A region of DNA containing four genes is shown, with the transcribed region of each gene highlighted in dark blue. The number of transcripts of each gene is indicated above the DNA (on a Y- axis). There are six transcripts of gene 1, one transcript of gene 2, twelve transcripts of gene 3, and no transcripts of gene 4. This is not an illustration of any actual set of genes and their transcription levels, but rather, illustrates that transcription is controlled individually for genes and other transcription units.

Translation (biology)

In molecular biology and genetics, translation is the process in which ribosomes in a cell'scytoplasm create proteins, following transcription of DNA to RNA in the cell's nucleus. The entire process is a part of gene expression.

In translation, messenger RNA (mRNA) is decoded by a ribosome, outside the nucleus, to produce a specific amino acid chain, or polypeptide. The polypeptide later folds into an active protein and performs its functions in the cell. The ribosome facilitates decoding by inducing the binding of complementary tRNA anticodon sequences to mRNA codons. The tRNAs carry specific amino acids that are chained together into a polypeptide as the mRNA passes through and is "read" by the ribosome.

Translation proceeds in three phases:

1. Initiation: The ribosome assembles around the target mRNA. The first tRNA is attached at the start codon.
2. Elongation: The tRNA transfers an amino acid to the tRNA corresponding to the next codon. The ribosome then moves (translocates) to the next mRNA codon to continue the process, creating an amino acid chain.
3. Termination: When a stop codon is reached, the ribosome releases the polypeptide.

In prokaryotes (bacteria), translation occurs in the cytoplasm, where the large and small subunits of the ribosome bind to the mRNA. In eukaryotes, translation occurs in the cytosol or across the membrane of the endoplasmic reticulum in a process called vectorial synthesis. In many instances, the entire ribosome/mRNA complex binds to the outer membrane of the rough endoplasmic reticulum (ER); the newly created polypeptide can be stored inside the ER for future vesicle transport and secretion outside of the cell, or immediately secreted.

Many types of transcribed RNA, such as transfer RNA, ribosomal RNA, and small nuclear RNA, do not undergo translation into proteins.

A number of antibiotics act by inhibiting translation. These include anisomycin, cycloheximide, chloramphenicol, tetracycline, streptomycin, erythromycin, and puromycin. Prokaryotic ribosomes have a different structure from that of eukaryotic ribosomes, and thus antibiotics can specifically target bacterial infections without any harm to a eukaryotic host's cells.

Basic mechanisms

The basic process of protein production is addition of one amino acid at a time to the end of a protein. This operation is performed by a ribosome. A ribosome is made up of two subunits, a small subunit and a large subunit. these subunits come together before translation of mRNA into a protein to provide a location for translation to be carried out and a polypeptide to be produced. The choice of amino acid type to add is determined by an mRNA molecule. Each amino acid added is matched to a three nucleotide subsequence of the mRNA. For each such triplet possible, the corresponding amino acid is accepted. The successive amino acids added to the chain are matched to successive nucleotide triplets in the mRNA. In this way the sequence of nucleotides in the template mRNA chain determines the sequence of amino acids in the generated amino acid chain. Addition of an amino acid occurs at the C-terminus of the peptide and thus translation is said to be amino-to-carboxyl directed.

The mRNA carries genetic information encoded as a ribonucleotide sequence from the chromosomes to the ribosomes. The ribonucleotides are "read" by translational machinery in a sequence of nucleotide triplets called codons. Each of those triplets codes for a specific amino acid.

The ribosome molecules translate this code to a specific sequence of amino acids. The ribosome is a multisubunit structure containing rRNA and proteins. It is the "factory" where amino acids are assembled into proteins. tRNAs are small noncoding RNA chains (74-93 nucleotides) that transport amino acids to the ribosome. tRNAs have a site for amino acid attachment, and a site called an anticodon. The anticodon is an RNA triplet complementary to the mRNA triplet that codes for their cargo amino acid.

Aminoacyl tRNA synthetases (enzymes) catalyze the bonding between specific tRNAs and the amino acids that their anticodon sequences call for. The product of this reaction is an aminoacyl-tRNA. This aminoacyl-tRNA is carried to the ribosome by EF-Tu, where mRNA codons are matched through complementary base pairing to specific tRNA anticodons. Aminoacyl-tRNA synthetases that mispair tRNAs with the wrong amino acids can produce mischarged aminoacyl-tRNAs, which can result in inappropriate amino acids at the respective position in protein. This "mistranslation" of the genetic code naturally occurs at low levels in most organisms, but certain cellular environments cause an increase in permissive mRNA decoding, sometimes to the benefit of the cell.

The ribosome has three sites for tRNA to bind. They are the aminoacyl site (abbreviated A), the peptidyl site (abbreviated P) and the exit site (abbreviated E). With respect to the mRNA, the three sites are oriented 5' to 3' E-P-A, because ribosomes move toward the 3' end of mRNA. The A site binds the incoming tRNA with the complementary codon on the mRNA. The P site holds the tRNA with the growing polypeptide chain. The E site holds the tRNA without its amino acid. When an aminoacyl-tRNA initially binds to its corresponding codon on the mRNA, it is in the A site. Then, a peptide bond forms between the amino acid of the tRNA in the A site and the amino acid of the charged tRNA in the P site. The growing polypeptide chain is transferred to the tRNA in the A site. Translocation occurs, moving the tRNA in the P site, now without an amino acid, to the E site; the tRNA that was in the A site, now charged with the polypeptide chain, is moved to the P site. The tRNA in the E site leaves and another aminoacyl-tRNA enters the A site to repeat the process.

After the new amino acid is added to the chain, and after the mRNA is released out of the nucleus and into the ribosome's core, the energy provided by the hydrolysis of a GTP bound to the translocase EF-G (in prokaryotes) and eEF-2 (in eukaryotes) moves the ribosome down one codon towards the 3' end. The energy required for translation of proteins is significant. For a protein containing n amino acids, the number of high-energy phosphate bonds required to translate it is 4n-1. The rate of translation varies; it is significantly higher in prokaryotic cells (up to 17-21 amino acid residues per second) than in eukaryotic cells (up to 6-9 amino acid residues per second).

Even though the ribosomes are usually considered accurate and processive machines, the translation process is subject to errors that can lead either to the synthesis of erroneous proteins or to the premature abandonment of translation. The rate of error in synthesizing proteins has been estimated to be between $1/10^5$ and $1/10^3$ misincorporated amino acids, depending on the experimental conditions. The rate of premature translation abandonment, instead, has been estimated to be of the order of magnitude of 10^{-4} events per translated codon. The correct amino acid is covalently bonded to the correct transfer RNA (tRNA) by amino acyl transferases. The amino acid is joined by its carboxyl group to the 3' OH of the tRNA by an ester bond. When the tRNA has an amino acid linked to it, the tRNA is termed "charged". Initiation involves the small subunit of the ribosome binding to the 5' end of mRNA with the help of initiation factors (IF). Termination of the polypeptide happens when the A site of the ribosome faces a stop codon (UAA, UAG, or UGA) on the mRNA. tRNA usually cannot recognize or bind to stop codons. Instead, the stop codon induces the binding of a release factor protein that prompts the disassembly of the entire ribosome/mRNA complex and the hydrolysis and the release of the polypeptide chain from the ribosome. Drugs or special sequence motifs on the mRNA can change the ribosomal structure so that near-cognate tRNAs are bound to the stop codon instead of the release factors. In such cases of 'translational readthrough', translation continues until the ribosome encounters the next stop codon.

The process of translation is highly regulated in both eukaryotic and prokaryotic organisms. Regulation of translation can impact the global rate of protein synthesis which is closely coupled to the metabolic and proliferative state of a cell. In addition, recent work has revealed that genetic differences and their subsequent expression as mRNAs can also impact translation rate in an RNA-specific manner.

Gene Expression and Regulation

Genetic code

Whereas other aspects such as the 3D structure, called tertiary structure, of protein can only be predicted using sophisticated algorithms, the amino acid sequence, called primary structure, can be determined solely from the nucleic acid sequence with the aid of a translation table.

This approach may not give the correct amino acid composition of the protein, in particular if unconventional amino acids such as selenocysteine are incorporated into the protein, which is coded for by a conventional stop codon in combination with a downstream hairpin (SEleno Cysteine Insertion Sequence, or SECIS).

There are many computer programs capable of translating a DNA/RNA sequence into a protein sequence. Normally this is performed using the Standard Genetic Code, however, few programs can handle all the "special" cases, such as the use of the alternative initiation codons. For instance, the rare alternative start codon CTG codes for Methionine when used as a start codon, and for Leucine in all other positions.

Translation tables

Even when working with ordinary eukaryotic sequences such as the Yeast genome, it is often desired to be able to use alternative translation tables—namely for translation of the mitochondrial genes. Currently the following translation tables are defined by the NCBI Taxonomy Group for the translation of the sequences in GenBank:

Post-transcriptional modification

Post-transcriptional modification or Co-transcriptional modification is the process in eukaryotic cells where primary transcript RNA is converted into mature RNA. A notable example is the conversion of precursor messenger RNA into mature messenger RNA (mRNA) that occurs prior to protein translation. The process includes three major steps: addition of a 5' cap, addition of a 3' poly-adenylation tail, and splicing. This process is vital for the correct translation of the genomes of eukaryotes because the initial precursor mRNA produced during transcription contains both exons (coding or important sequences involved in translation), and introns (non-coding sequences).

Actually Post-transcriptional modification or Co-transcriptional modification is the process in eukaryotic cells where primary transcript RNA is converted into mature RNA. A notable example is the conversion of precursor messenger RNA into mature messenger RNA (mRNA) that occurs prior to protein translation.

In eukaryotes, a synthesized RNA transcript may undergo a number of levels of processing known as post-transcriptional modifications. In the case of messenger RNA, modification is necessary to convert pre-mRNA into a mature mRNA that is ready for protein translation.

One such modification is splicing. The RNA contains a mixture of introns and exons when it is first transcribed. Splicing serves to remove introns, and it occurs either by the action of a spliceosome, or by self-splicing. Alternative splicing shuffles the composition of exons in a transcript, meaning that numerous different transcripts can be made from the same gene.

In mRNA, the 3' end is modified by the addition of multiple adenine residues to produce a structure called a poly-adenylated (poly A) tail. The length of a poly-A tail is highly variable, but over the lifetime of the mRNA these adenine residues are slowly removed. Thus the length of a poly-A tail determines the half-life of mRNA: a longer tail means a longer half-life. The poly-A tail also makes it easier for us to purify eukaryotic mRNA in the lab by annealment to a complementary poly-thymine (poly-T) primer.

In mRNA, the 5' end is modified by the addition of a 7-methylguanosine cap (a methylated guanine residue). The cap is attached the 'wrong way round' forming a 5'-5' pyrophosphate bond, rather than the usual 3'-5' phosphodiester bond. This has two key roles: it enables ribosomal recognition, and thus greater efficiency of protein translation, as well as protecting the mRNA from degradation by 5'-3' exonuclease enzymes.

Note that the 5' cap is the 'front' end of the mRNA and the 3' poly-A tail is the 'back' end. The coding sequence of the transcript lies in between.

After the 5' cap, but before the initation AUG codon, is a 5' untranslated region (UTR) or leader sequence. Equally, after the termination codon, but before the 3' untranslated region (UTR) or trailer sequence is a 3' untranslated region (UTR) or trailer sequence. These are not protein-coding but may have regulatory roles; for instance, the 3' UTR is known to act as a poly-adenylation signal for the addition of the poly-A tail by various enzymes.

Flow of genetic information is relatively simple and fast in prokaryotes where both transcription and translation are coupled in the protoplasmic compartment. But in eukaryotes the nuclear envelop is a hurdle in this smooth flow with transcription taking place in the nuclear compartment and translation in the cytoplasmic compartment. Besides, the product of transcription in eukaryotes is a raw transcript (called primary transcript or pre-RNA) which undergoes various chemical changes to become a functional molecule. The term RNA processing refers to all these chemical modifications necessary to generate a final RNA product from the primary transcript. Processing typically involves the addition or chemical modifications of specific nucleotides and also the removal of certain portions of precursor RNA (Table 5.1). Although processing is more often related to the eukaryotic system, nevertheless prokaryotic rRNAs and tRNAs (not mRNAs) are also synthesized as large precursor molecules which need to be processed to yield mature RNAs. All the three major classes of RNA (mRNA, rRNA and tRNA) in eukaryotes undergo varying degrees of modifications in the nucleus prior to their export into the cytoplasm where they are utilised for translating the genetic information into a functional product, the protein.

The number and size of introns varies from species to species and from gene to gene [Table 2]. For example only about 200 of the 6300 genes in yeast have introns and out of these 200 intron containing genes most have only one intron. In contrast the number of introns in human varies from none in a few genes and two in a ß-Globin gene to as many as 363 in titin gene. The average size of a mammalian intron is about 2000 nucleotides and that of exon is about 200 nucleotides; thus exons account for only about 10% of total length of the gene. An extreme example is the gene for human Duchenne muscular dystrophy which is about 2.4 million base pairs long that produces an mRNA of about 14000 nucleotides. This gene has 78 introns representing more than 99% of the gene's DNA!

The absence of introns from mature mRNA raises the question of whether the introns present in DNA are actually transcribed into the primary transcript (pre-mRNA). This question has been addressed by experiments in which single stranded RNA is hybridized to double stranded DNA under conditions that favour the formation of hybrids between complementary regions of RNA and DNA and these were examined by electron microscopy. containing alternating blocks of exons and introns and during the processing of pre-mRNA these introns are removed.

At the time introns were discovered, it was known that the nucleus contains a large population of RNA molecules of various sizes (size ranges from 2000-20000 bases). These RNA molecules are called heterogeneous nuclear RNA or hnRNA. They all are RNA polymerase II transcripts and have a very short half life (of the order of minutes). Most of them are capped at the 5' end and polyadenylated at the 3' end. It is now clear that these hnRNAs are precursors to cytoplasmic mRNA. But these nuclear hnRNA are much larger in size than cytoplasmic mRNA. This difference in size is due to the presence of introns within the coding sequence in mRNA precursors. Much of the hnRNA actually consist of pre-mRNA molecules in various stages of processing before they leave the nucleus. Introns are not exclusive to protein coding genes of eukaryotes, however. They have been found in tRNA genes, in rRNA genes, in mitochondrial and chloroplast genes as well as in few genes of bacteriophages. In fact the existence of introns was first observed in adenovirus (a DNA virus that infects humans) by two different groups of scientists working independently in the laboratories of Richard Roberts and Philip Sharp in 1976. Splicing mechanisms The discovery of introns in genes generated interest in finding out the mechanisms by which these intron sequences are prevented from being expressed in the protein product. The demonstration that RNA polymerase transcribes intron sequences along with the exon sequences and ribosomes do not skip any codon while translating the messenger RNA finally focussed the research on removal of introns from the primary gene transcript. This entire process of removal of introns from the precursor RNA and joining of exons to form functional RNA is called RNA splicing.

There are three distinct mechanisms of intron excision from RNA transcripts. The introns of nuclear pre-mRNAs are spliced out by complex ribonucleoprotein particles called spliceosomes. 2. The introns of some rRNA precursors are removed autocatalytically by the RNA molecule itself. 3. The introns of tRNA precursors are excised by enzymes – splicing endonuclease and splicing ligase. These three mechanisms of intron excision are discussed in the following sections. Spliceosomes remove introns from pre-mRNA To produce a functional mRNA from a pre-mRNA in eukaryotes, three kinds of posttranscriptional modifications take place. These are- addition of 7-methyl guanosine cap at 5'end, addition of Poly A tail at 3'end and removal of introns. Here we describe the third type of modification i.e. removal of introns by RNA splicing. To splice an RNA, breaks in the strand must be introduced at the 5' and 3' edges of the intron (The intron-exon boundary at the 5'end of intron is called 5' splice site and that at the 3'end of intron is called 3' splice site) and the exons on either side of the splice sites must be covalently joined. Splicing must occur with great precision because the addition or loss of even a single nucleotide would alter the mRNA reading frame and result in mistranslation.

The genetic code is frequently referred to as a "blueprint" because it contains the instructions a cell requires in order to sustain itself. We now know that there is more to these instructions than simply the sequence of letters in the nucleotide code, however. For example, vast amounts of evidence demonstrate that this code is the basis for the production of various molecules, including RNA and protein. Research has also shown that the instructions stored within DNA are "read" in two steps: transcription and translation. In transcription, a portion of the double-stranded DNA template gives rise to a single-stranded RNA molecule. In some cases, the RNA molecule itself is a "finished product" that serves some important function within the cell. Often, however, transcription of an RNA molecule is followed by a translation step, which ultimately results in the production of a protein molecule.

Gene Expression and Regulation

Visualizing Transcription

The process of transcription can be visualized by electron microscopy (Figure 1); in fact, it was first observed using this method in 1970. In these early electron micrographs, the DNA molecules appear as "trunks," with many RNA "branches" extending out from them. When DNAse and RNAse (enzymes that degrade DNA and RNA, respectively) were added to the molecules, the application of DNAse eliminated the trunk structures, while the use of RNAse wiped out the branches.

DNA is double-stranded, but only one strand serves as a template for transcription at any given time. This template strand is called the noncoding strand. The nontemplate strand is referred to as the coding strand because its sequence will be the same as that of the new RNA molecule. In most organisms, the strand of DNA that serves as the template for one gene may be the nontemplate strand for other genes within the same chromosome.

The Transcription Process

The process of transcription begins when an enzyme called RNA polymerase (RNA pol) attaches to the template DNA strand and begins to catalyze production of complementary RNA. Polymerases are large enzymes composed of approximately a dozen subunits, and when active on DNA, they are also typically complexed with other factors. In many cases, these factors signal which gene is to be transcribed.

Three different types of RNA polymerase exist in eukaryotic cells, whereas bacteria have only one. In eukaryotes, RNA pol I transcribes the genes that encode most of the ribosomal RNAs (rRNAs), and RNA pol III transcribes the genes for one small rRNA, plus the transfer RNAs that play a key role in the translation process, as well as other small regulatory RNA molecules. Thus, it is RNA pol II that transcribes the messenger RNAs, which serve as the templates for production of protein molecules.

Transcription Initiation

The first step in transcription is initiation, when the RNA pol binds to the DNA upstream (5') of the gene at a specialized sequence called a promoter (Figure 2a). In bacteria, promoters are usually composed of three sequence elements, whereas in eukaryotes, there are as many as seven elements.

In prokaryotes, most genes have a sequence called the Pribnow box, with the consensus sequence TATAAT positioned about ten base pairs away from the site that serves as the location of transcription initiation. Not all Pribnow boxes have this exact nucleotide sequence; these nucleotides are simply the most common ones found at each site. Although substitutions do occur, each box nonetheless resembles this consensus fairly closely. Many genes also have the consensus sequence TTGCCA at a position 35 bases upstream of the start site, and some have what is called an upstream element, which is an A-T rich region 40 to 60 nucleotides upstream that enhances the rate of transcription (Figure 3). In any case, upon binding, the RNA pol "core enzyme" binds to another subunit called the sigma subunit to form a holoezyme capable of unwinding the DNA double helix in order to facilitate access to the gene. The sigma subunit conveys promoter specificity to RNA polymerase; that is, it is responsible for telling RNA polymerase where to bind. There are a number of different sigma subunits that bind to different promoters and therefore assist in turning genes on and off as conditions change.

Eukaryotic promoters are more complex than their prokaryotic counterparts, in part because eukaryotes have the aforementioned three classes of RNA polymerase that transcribe different sets of genes. Many eukaryotic genes also possess enhancer sequences, which can be found at considerable distances from the genes they affect. Enhancer sequences control gene activation by binding with activator proteins and altering the 3-D structure of the DNA to help "attract" RNA pol II, thus regulating transcription. Because eukaryotic DNA is tightly packaged as chromatin, transcription also requires a number of specialized proteins that help make the template strand accessible.

In eukaryotes, the "core" promoter for a gene transcribed by pol II is most often found immediately upstream (5') of the start site of the gene. Most pol II genes have a TATA box (consensus sequence TATTAA) 25 to 35 bases upstream of the initiation site, which affects the transcription rate and determines location of the start site. Eukaryotic RNA polymerases use a number of essential cofactors (collectively called general transcription factors), and one of these, TFIID, recognizes the TATA box and ensures that the correct start site is used. Another cofactor, TFIIB, recognizes a different common consensus sequence, G/C G/C G/C G C C C, approximately 38 to 32 bases upstream.

The terms "strong" and "weak" are often used to describe promoters and enhancers, according to their effects on transcription rates and thereby on gene expression. Alteration of promoter strength can have deleterious effects upon a cell, often resulting in disease. For example, some tumor-promoting viruses transform healthy cells by inserting strong promoters in the vicinity of growth-stimulating genes, while translocations in some cancer cells place genes that should be "turned off" in the proximity of strong promoters or enhancers.

Enhancer sequences do what their name suggests: They act to enhance the rate at which genes are transcribed, and their effects can be quite

Gene Expression and Regulation

powerful. Enhancers can be thousands of nucleotides away from the promoters with which they interact, but they are brought into proximity by the looping of DNA. This looping is the result of interactions between the proteins bound to the enhancer and those bound to the promoter. The proteins that facilitate this looping are called activators, while those that inhibit it are called repressors.

Transcription of eukaryotic genes by polymerases I and III is initiated in a similar manner, but the promoter sequences and transcriptional activator proteins vary.

Strand Elongation

Once transcription is initiated, the DNA double helix unwinds and RNA polymerase reads the template strand, adding nucleotides to the 3' end of the growing chain (Figure 2b). At a temperature of 37 degrees Celsius, new nucleotides are added at an estimated rate of about 42-54 nucleotides per second in bacteria (Dennis & Bremer, 1974), while eukaryotes proceed at a much slower pace of approximately 22-25 nucleotides per second (Izban & Luse, 1992).

Transcription Termination

Terminator sequences are found close to the ends of noncoding sequences (Figure 2c). Bacteria possess two types of these sequences. In rho-independent terminators, inverted repeat sequences are transcribed; they can then fold back on themselves in hairpin loops, causing RNA pol to pause and resulting in release of the transcript (Figure 5). On the other hand, rho-dependent terminators make use of a factor called rho, which actively unwinds the DNA-RNA hybrid formed during transcription, thereby releasing the newly synthesized RNA.

In eukaryotes, termination of transcription occurs by different processes, depending upon the exact polymerase utilized. For pol I genes, transcription is stopped using a termination factor, through a mechanism similar to rho-dependent termination in bacteria. Transcription of pol III genes ends after transcribing a termination sequence that includes a polyuracil stretch, by a mechanism resembling rho-independent prokaryotic termination. Termination of pol II transcripts, however, is more complex.

Transcription of pol II genes can continue for hundreds or even thousands of nucleotides beyond the end of a noncoding sequence. The RNA strand is then cleaved by a complex that appears to associate with the polymerase. Cleavage seems to be coupled with termination of transcription and occurs at a consensus sequence. Mature pol II mRNAs are polyadenylated at the 3'-end, resulting in a poly(A) tail; this process follows cleavage and is also coordinated with termination.

Both polyadenylation and termination make use of the same consensus sequence, and the interdependence of the processes was demonstrated in the late 1980s by work from several groups. One group of scientists working with mouse globin genes showed that introducing mutations into the consensus sequence AATAAA, known to be necessary for poly(A) addition, inhibited both polyadenylation and transcription termination. They measured the extent of termination by hybridizing transcripts with the different poly(A) consensus sequence mutants with wild-type transcripts, and they were able to see a decrease in the signal of hybridization, suggesting that proper termination was inhibited. They therefore concluded that polyadenylation was necessary for termination (Logan et. al., 1987). Another group obtained similar results using a monkey viral system, SV40 (simian virus 40). They introduced mutations into a poly(A) site, which caused mRNAs to accumulate to levels far above wild type (Connelly & Manley, 1988).

The exact relationship between cleavage and termination remains to be determined. One model supposes that cleavage itself triggers termination; another proposes that polymerase activity is affected when passing through the consensus sequence at the cleavage site, perhaps through changes in associated transcriptional activation factors. Thus, research in the area of prokaryotic and eukaryotic transcription is still focused on unraveling the molecular details of this complex process, data that will allow us to better understand how genes are transcribed and silenced.

Regulation of Gene Expression in Eukaryotes

In eukaryotes, the expression of biologically active proteins can be modulated at several points as follows:

Chromatin Structure

Eukaryotic DNA is compacted into chromatin structures which can be altered by histone modifications. Such modifications can result in the up- or down-regulation of a gene.

Initiation of Transcription

This is a key point of regulation of eukaryotic gene expression. Here, several factors such as promoters and enhancers alter the ability of RNA polymerase to transcribe the mRNA, thus modulating the expression of the gene.

Post-Transcriptional Processing

Modifications such as polyadenylation, splicing, and capping of the pre-mRNA transcript in eukaryotes can lead to different levels and patterns of gene expression. For example, different splicing patterns for the same gene will generate biologically different proteins following translation.

RNA Transport

After post-transcriptional processing, the mature mRNA must be transported from the nucleus to the cytosol so that it can be translated into a protein. This step is a key point of regulation of gene expression in eukaryotes.

Stability of mRNAs

Eukaryotic mRNAs differ in their stability and some unstable transcripts usually have sequences that bind to microRNAs and reduce the stability of mRNAs, resulting in down-regulation of the corresponding proteins.

Initiation of Translation

At this stage, the ability of ribosomes in recognizing the start codon can be modulated, thus affecting the expression of the gene. Several examples of translation initiation at non-AUG codons in eukaryotes are available.

Post-Translational Processing

Common modifications in polypeptide chains include glycosylation, fatty acylation, and acetylation - these can help in regulating expression of the gene and offering vast functional diversity.

Protein Transport and Stability

Following translation and processing, proteins must be carried to their site of action in order to be biologically active. Also, by controlling the stability of proteins, the gene expression can be controlled. Stability varies greatly depending on specific amino acid sequences present in the proteins.

REGULATION OF GENE EXPRESSION IN PROKARYOTES

Prokaryotic genes are clustered into operons, each of which code for a corresponding protein.

In prokaryotes, transcription initiation is the main point of control of gene expression. It is chiefly controlled by 2 DNA sequence elements of size 35 bases and 10 bases, respectively. These elements are called promoter sequences as they help RNA polymerase recognize the start sites of transcription. RNA polymerase recognizes and binds to these promoter sequences. The interaction of RNA polymerase with promoter sequences is in turn controlled by regulatory proteins called activators or repressors based on whether they positively or negatively affect the recognition of promoter sequence by RNA pol.

There are 2 major modes of transcriptional control in E. coli to modulate gene expression. Both of these control mechanisms involve repressor proteins.

Catabolite-Regulated

In this system, control is exerted upon operons that produce genes necessary for the energy utilization. The lac operon is an example of this in E. coli.

In E. coli, glucose has a positive effect on the expression of genes that encode enzymes involved in the catabolism of alternative sources of carbon and energy such as lactose. Due to the preference for glucose, in its presence enzymes involved in the catabolism of other energy sources are not expressed. In this way, glucose represses the lac operon even if an inducer (lactose) is present.

Transcriptional Attenuation

This modulates operons necessary for biomolecule synthesis. This is called attenuated operon as the operons are attenuated by specific sequences present in the transcribed RNA – gene expression is therefore dependent on the ability of RNA Polymerase to continue elongation past specific sequences. An example of an attenuated operon is the trp operon which encodes five enzymes necessary for tryptophan biosynthesis in E.coli. These genes are expressed only when tryptophan synthesis is necessary i.e. when tryptophan is not environmentally present. This is partly controlled when a repressor binds to tryptophan and prevents transcription for unnecessary tryptophan biosynthesis.

Relationship of cell growth to the regulation of tissue-specific gene expression during osteoblast differentiation.

The relationship of cell proliferation to the temporal expression of genes characterizing a developmental sequence associated with bone cell differentiation can be examined in primary diploid cultures of fetal calvarial-derived osteoblasts by the combination of molecular, biochemical, histochemical, and ultrastructural approaches. Modifications in gene expression define a developmental sequence that has 1) three principal periods: proliferation, extracellular matrix maturation, and mineralization; and 2) two restriction points to which the cells can progress but cannot pass without further signals. The first restriction point is when proliferation is down-regulated and gene expression associated with extracellular matrix maturation is induced, and the second when mineralization occurs. Initially, actively proliferating cells, expressing cell cycle and cell growth regulated

Gene Expression and Regulation

genes, produce a fibronectin/type I collagen extracellular matrix. A reciprocal and functionally coupled relationship between the decline in proliferative activity and the subsequent induction of genes associated with matrix maturation and mineralization is supported by 1) a temporal sequence of events in which an enhanced expression of alkaline phosphatase occurs immediately after the proliferative period, and later an increased expression of osteocalcin and osteopontin at the onset of mineralization; 2) increased expression of a specific subset of osteoblast phenotype markers, alkaline phosphatase and osteopontin, when proliferation is inhibited; and 3) enhanced levels of expression of the osteoblast markers when collagen deposition is promoted, suggesting that the extracellular matrix contributes to both the shutdown of proliferation and development of the osteoblast phenotype. The loss of stringent growth control in transformed osteoblasts and in osteosarcoma cells is accompanied by a deregulation of the tightly coupled relationship between proliferation and progressive expression of genes associated with bone cell differentiation.

In the post Human Genome Project era there are efforts to understand translation of the genomic sequence into the transcriptome. The human transcriptome is represented by >100,000 distinct transcripts presently described for ~20,000 protein-coding genes. Additionally, mRNA isoforms are produced by alternative processing of primary RNA transcripts. This alternative splicing affects >90% of the human genes and has been suggested to the primary driver of phenotypic complexity. Despite this diversity in the coding sequence, the nonprotein- coding molecules contribute to > 95% of the transcriptome. Numerous technologies and experimental platforms have facilitated the investigation of the complexity in the transcriptome that greater than that of the simple genome sequence.

The recent RNA sequencing or RNA-Seq, involves highthroughput sequencing of short cDNA fragments obtained from the pool of RNA (total or fractionated, such as poly(A)+ or ribosomal rna depleted) to provide single-base resolution to the transcriptome. The traditional expression analysis primarily addressed the identification of differentially expressed transcripts with respect to measured variables of interest, such as differing environments, treatments, phenotypes, or clinical outcomes. The advent of RNA-Seq has provided a broad spectrum of applications and enabled researchers to address a wider range of biological problems. This technology enables cataloguing all species of transcript, including coding and non-coding mRNAs; to determine the transcriptional structure of genes, splicing patterns and other post-transcriptional modifications.

Despite, the breadth of possibilities RNA-Seq measurements and analysis of expression remains a field of active research. The major concerns and scrutiny is attributed to the numerous technical and analytical limitations. Early concerns regarding library preparation, sequencing error, read mapping, and gene expression quantification have been resolved by a number of studies; however, there is no standardized approach for quality control and data adjustment of RNASeq data after the generation of gene expression estimates. As a caution, without an appropriate approach to data analysis, reproducibility of these studies remains limited. There are numerous studies that are providing frameworks and strategies to assess possible sample contamination and assess the biologic validity of each data analysis step to ultimately enable confident downstream analyses.

An important consideration in gene expression is still the biological source for RNA profiling. To elaborate for disease relevant questions there is a clear and compelling need to conduct gene expression studies in tissues that are specifically relevant to the disease of interest as opposed to cell lines. It is reassuring that studies have reported that robust gene expression can be obtained using RNA from autopsy-derived tissue 24 hours after autolysis [5]. However, examination of a tissue which is a heterogeneous mix of several distinct cell populations makes it difficult to distinguish whether gene expression variability reflects shifts in cell proportions or variable cell-type specific expression.

In addition to cell-type variability, gene expression data is also confounded by various known and unknown sources of variation such as batch effects, environmental influences and sample history. These unknown confounders that plague comparative expression analysis are not easily attributable to any recorded measurement. These unknown covariates can be approximated through various data decomposition methods, like Principal Components Analysis (PCA), Surrogate Variable Analysis (SVA), Independent Surrogate Variable Analysis (ISVA) and Probabilistic Estimation of Expression Residuals (PEER). This method can be used to correct for biases and provide accurate estimates of global comparison of gene analysis and for detecting genetic associations in expression data (eQTL).

Although, there are suggestions to correct the variability of expression that may be caused by difference in cell-type proportions. Studies have reported that underlying mechanism in some human diseases are accompanied by changes of cell populations in corresponding tissues. There are computational methods of analyzing gene expression in samples of varying composition that can improve analyses of quantitative molecular data in many biological contexts. There is also the development of sequencing-based technologies that are increasingly being targeted to individual cells, which will allow many new and longstanding questions to be addressed.

The maturation of single-cell transcriptomics should provide indepth knowledge of the precise transcript map and the regulatory landscape in individual single cells at different levels of resolution (single cell, cell-type or tissue). This level of resolution should be beneficial towards insightful biomarker discovery and disease diagnostics.

Regulation of gene expression

Gene expression is the process by which information from a gene is used in the synthesis of a functional gene product. These products are often proteins, but in non-protein coding genes such as transfer RNA (tRNA) or small nuclear RNA (snRNA) genes, the product is a functional RNA. The process of gene expression is used by all known life—eukaryotes (including multicellular organisms), prokaryotes (bacteria and archaea), and utilized by viruses—to generate the macromolecular machinery for life.

Several steps in the gene expression process may be modulated, including the transcription, RNA splicing, translation, and post-translational modification of a protein. Gene regulation gives the cell control over structure and function, and is the basis for cellular differentiation, morphogenesis and the versatility and adaptability of any organism. Gene regulation may also serve as a substrate for evolutionary change, since control of the timing, location, and amount of gene expression can have a profound effect on the functions (actions) of the gene in a cell or in a multicellular organism.

In genetics, gene expression is the most fundamental level at which the genotype gives rise to the phenotype, i.e. observable trait. The genetic code stored in DNA is "interpreted" by gene expression, and the properties of the expression give rise to the organism's phenotype. Such phenotypes are often expressed by the synthesis of proteins that control the organism's shape, or that act as enzymes catalysing specific metabolic pathways characterising the organism. Regulation of gene expression is thus critical to an organism's development.

Regulation of gene expression refers to the control of the amount and timing of appearance of the functional product of a gene. Control of expression is vital to allow a cell to produce the gene products it needs when it needs them; in turn, this gives cells the flexibility to adapt to a variable environment, external signals, damage to the cell, and other stimuli. More generally, gene regulation gives the cell control over all structure and function, and is the basis for cellular differentiation, morphogenesis and the versatility and adaptability of any organism.

Numerous terms are used to describe types of genes depending on how they are regulated; these include:

- A constitutive gene is a gene that is transcribed continually as opposed to a facultative gene, which is only transcribed when needed.
- A housekeeping gene is a gene that is required to maintain basic cellular function and so is typically expressed in all cell types of an organism. Examples include actin, GAPDH and ubiquitin. Some housekeeping genes are transcribed at a relatively constant rate and these genes can be used as a reference point in experiments to measure the expression rates of other genes.
- A facultative gene is a gene only transcribed when needed as opposed to a constitutive gene.
- An inducible gene is a gene whose expression is either responsive to environmental change or dependent on the position in the cell cycle.

Any step of gene expression may be modulated, from the DNA-RNA transcription step to post-translational modification of a protein. The stability of the final gene product, whether it is RNA or protein, also contributes to the expression level of the gene—an unstable product results in a low expression level. In general gene expression is regulated through changes in the number and type of interactions between molecules that collectively influence transcription of DNA and translation of RNA.

Some simple examples of where gene expression is important are:
- Control of insulin expression so it gives a signal for blood glucose regulation.
- X chromosome inactivation in female mammals to prevent an "overdose" of the genes it contains.
- Cyclin expression levels control progression through the eukaryotic cell cycle.

The control of gene expression is a biological process essential to all organisms. This is accomplished through the interaction of regulatory proteins with specific DNA motifs in the control regions of the genes that they regulate. Upon binding to DNA, and through specific protein-protein interactions, these regulatory proteins convey signals to the basal transcriptional machinery, containing the respective RNA polymerases, resulting in particular rates of gene expression. In eukaryotes, in addition and complementary to the binding of regulatory proteins to DNA, chromatin structure plays a role in modulating gene expression. Small RNAs are emerging as key components in this process. This chapter provides an introduction to some of the basic players participating in these processes, the transcription factors and co-regulators, the cis-regulatory elements that often function as transcription factor docking sites, and the emerging role of small RNAs in the regulation of gene expression.

Gene Expression and Regulation   183

Key points:

- Gene regulation is the process of controlling which genes in a cell's DNA are expressed (used to make a functional product such as a protein).
- Different cells in a multicellular organism may express very different sets of genes, even though they contain the same DNA.
- The set of genes expressed in a cell determines the set of proteins and functional RNAs it contains, giving it its unique properties.
- In eukaryotes like humans, gene expression involves many steps, and gene regulation can occur at any of these steps. However, many genes are regulated primarily at the level of transcription.

Your amazing body contains hundreds of different cell types, from immune cells to skin cells to neurons. Almost all of your cells contain the same set of DNA instructions – so why do they look so different, and do such different jobs? The answer: different gene regulation!

Gene regulation makes cells different

Gene regulation is how a cell controls which genes, out of the many genes in its genome, are "turned on" (expressed). Thanks to gene regulation, each cell type in your body has a different set of active genes – despite the fact that almost all the cells of your body contain the exact same DNA. These different patterns of gene expression cause your various cell types to have different sets of proteins, making each cell type uniquely specialized to do its job.

For example, one of the jobs of the liver is to remove toxic substances like alcohol from the bloodstream. To do this, liver cells express genes encoding subunits (pieces) of an enzyme called alcohol dehydrogenase. This enzyme breaks alcohol down into a non-toxic molecule. The neurons in a person's brain don't remove toxins from the body, so they keep these genes unexpressed, or "turned off." Similarly, the cells of the liver don't send signals using neurotransmitters, so they keep neurotransmitter genes turned off.

Left panel: liver cell. The liver cell contains alcohol dehydrogenase proteins. If we look in the nucleus, we see that an alcohol dehydrogenase gene is expressed to make RNA, while a neurotransmitter gene is not. The RNA is processed and translated, which is why the alcohol dehydrogenase proteins are found in the cell.

Right panel: neuron. The neuron contains neurotransmitter proteins. If we look in the nucleus, we see that the alcohol dehydrogenase gene is not expressed to make RNA, while the neurotransmitter gene is. The RNA is processed and translated, which is why the neurotransmitter proteins are found in the cell.

There are many other genes that are expressed differently between liver cells and neurons (or any two cell types in a multicellular organism like yourself).

How do cells "decide" which genes to turn on?

Now there's a tricky question! Many factors that can affect which genes a cell expresses. Different cell types express different sets of genes, as we saw above. However, two different cells of the same type may also have different gene expression patterns depending on their environment and internal state.

Broadly speaking, we can say that a cell's gene expression pattern is determined by information from both inside and outside the cell.

- Examples of information from inside the cell: the proteins it inherited from its mother cell, whether its DNA is damaged, and how much ATP it has.
- Examples of information from outside the cell: chemical signals from other cells, mechanical signals from the extracellular matrix, and nutrient levels.

How do these cues help a cell "decide" what genes to express? Cells don't make decisions in the sense that you or I would. Instead, they have molecular pathways that convert information – such as the binding of a chemical signal to its receptor – into a change in gene expression.

As an example, let's consider how cells respond to growth factors. A growth factor is a chemical signal from a neighboring cell that instructs a target cell to grow and divide. We could say that the cell "notices" the growth factor and "decides" to divide, but how do these processes actually occur?

Growth factors bind to their receptors on the cell surface and activate a signaling pathway in the cell. The signaling pathway activates transcription factors in the nucleus, which bind to DNA near division-promoting and growth-promoting genes and cause them to be transcribed into RNA. The RNA is processed and exported from the nucleus, then translated to make proteins that drive growth and division.

- The cell detects the growth factor through physical binding of the growth factor to a receptor protein on the cell surface.
- Binding of the growth factor causes the receptor to change shape, triggering a series of chemical events in the cell that activate proteins called transcription factors.
- The transcription factors bind to certain sequences of DNA in the nucleus and cause transcription of cell division-related genes.

Gene Expression and Regulation 185

- The products of these genes are various types of proteins that make the cell divide (drive cell growth and/or push the cell forward in the cell cycle).

This is just one example of how a cell can convert a source of information into a change in gene expression. There are many others, and understanding the logic of gene regulation is an area of ongoing research in biology today.

Growth factor signaling is complex and involves the activation of a variety of targets, including both transcription factors and non-transcription factor proteins. You can learn more about how growth factor signaling works in the article on intracellular signal transduction.

EUKARYOTIC GENE EXPRESSION CAN BE REGULATED AT MANY STAGES

In the articles that follow, we'll examine different forms of eukaryotic gene regulation. That is, we'll see how the expression of genes in eukaryotes (like us!) can be controlled at various stages, from the availability of DNA to the production of mRNAs to the translation and processing of proteins.

Eukaryotic gene expression involves many steps, and almost all of them can be regulated. Different genes are regulated at different points, and it's not uncommon for a gene (particularly an important or powerful one) to be regulated at multiple steps.

- Chromatin accessibility. The structure of chromatin (DNA and its organizing proteins) can be regulated. More open or "relaxed" chromatin makes a gene more available for transcription.
- Transcription. Transcription is a key regulatory point for many genes. Sets of transcription factor proteins bind to specific DNA sequences in or near a gene and promote or repress its transcription into an RNA.
- RNA processing. Splicing, capping, and addition of a poly-A tail to an RNA molecule can be regulated, and so can exit from the nucleus. Different mRNAs may be made from the same pre-mRNA by alternative splicing.

Stages of eukaryotic gene expression (any of which can be potentially regulated).

1. Chromatin structure. Chromatin may be tightly compacted or loose and open.
2. Transcription. An available gene (with sufficiently open chromatin) is transcribed to make a primary transcript.
3. Processing and export. The primary transcript is processed (spliced, capped, given a poly-A tail) and shipped out of the nucleus.

4. mRNA stability. In the cytosol, the mRNA may be stable for long periods of time or may be quickly degraded (broken down).
5. Translation. The mRNA may be translated more or less readily/frequently by ribosomes to make a polypeptide.
6. Protein processing. The polypeptide may undergo various types of processing, including proteolytic cleavage (snipping off of amino acids) and addition of chemical modifications, such as phosphate groups.

All these steps (if applicable) need to be executed for a given gene for an active protein to be present in the cell.

- RNA stability. The lifetime of an mRNA molecule in the cytosol affects how many proteins can be made from it. Small regulatory RNAs called miRNAs can bind to target mRNAs and cause them to be chopped up.
- Translation. Translation of an mRNA may be increased or inhibited by regulators. For instance, miRNAs sometimes block translation of their target mRNAs (rather than causing them to be chopped up).
- Protein activity. Proteins can undergo a variety of modifications, such as being chopped up or tagged with chemical groups. These modifications can be regulated and may affect the activity or behavior of the protein.

Although all stages of gene expression can be regulated, the main control point for many genes is transcription. Later stages of regulation often refine the gene expression patterns that are "roughed out" during transcription.

To learn more, see the articles on transcription factors and regulation after transcription.

Gene regulation and differences between species

Differences in gene regulation makes the different cell types in a multicellular organism (such as yourself) unique in structure and function. If we zoom out a step, gene regulation can also help us explain some of the differences in form and function between different species with relatively similar gene sequences.

For instance, humans and chimpanzees have genomes that are about 98.8\%98.8%98, point, 8, percentidentical at the DNA level. The protein-coding sequences of some genes are different between humans and chimpanzees, contributing to the differences between the species. However, researchers also think that changes in gene regulation play a major role in making humans and chimps different from one another. For instance, some DNA regions that are present in the chimpanzee genome but missing in the

Gene Expression and Regulation

human genome contain known gene-regulatory sequences that control when, where, or how strongly a gene is expressed[3].

Conclusion

How does a gene, which consists of a string of DNA hidden in a cell's nucleus, know when it should express itself? How does this gene cause the production of a string of amino acids called a protein? How do different types of cells know which types of proteins they must manufacture? The answers to such questions lie in the study of gene expression. Thus, this collection or articles begins by showing how a quiet, well-guarded string of DNA is expressed to make RNA, and how the messenger RNA is translated from nucleic acid coding to protein coding to form a protein. Along the way, the article set also examines the nature of the genetic code, how the elements of code were predicted, and how the actual codons were determined.

Next, we turn to the regulation of genes. Genes can't control an organism on their own; rather, they must interact with and respond to the organism's environment. Some genes are constitutive, or always "on," regardless of environmental conditions. Such genes are among the most important elements of a cell's genome, and they control the ability of DNA to replicate, express itself, and repair itself. These genes also control protein synthesis and much of an organism's central metabolism. In contrast, regulated genes are needed only occasionally — but how do these genes get turned "on" and "off"? What specific molecules control when they are expressed?

It turns out that the regulation of such genes differs between prokaryotes and eukaryotes. For prokaryotes, most regulatory proteins are negative and therefore turn genes off. Here, the cells rely on protein–small molecule binding, in which a ligand or small molecule signals the state of the cell and whether gene expression is needed. The repressor or activator protein binds near its regulatory target: the gene. Some regulatory proteins must have a ligand attached to them to be able to bind, whereas others are unable to bind when attached to a ligand. In prokaryotes, most regulatory proteins are specific to one gene, although there are a few proteins that act more widely. For instance, some repressors bind near the start of mRNA production for an entire operon, or cluster of coregulated genes. Furthermore, some repressors have a fine-tuning system known as attenuation, which uses mRNA structure to stop both transcription and translation depending on the concentration of an operon's end-product enzymes. (In eukaryotes, there is no exact equivalent of attenuation, because transcription occurs in the nucleus and translation occurs in the cytoplasm, making this sort of coordinated effect impossible.) Yet another layer of prokaryotic regulation affects the structure of RNA polymerase, which turns on large groups of genes. Here, the sigma factor of RNA polymerase changes several times to produce heat- and desiccation-resistant spores. Here, the articles on prokaryotic regulation delve into each of these topics, leading to primary literature in many cases.

For eukaryotes, cell-cell differences are determined by expression of different sets of genes. For instance, an undifferentiated fertilized egg looks and acts quite different from a skin cell, a neuron, or a muscle cell because of differences in the genes each cell expresses. A cancer cell acts different from a normal cell for the same reason: It expresses different genes. (Using microarray analysis, scientists can use such differences to assist in diagnosis and selection of appropriate cancer treatment.) Interestingly, in eukaryotes, the default state of gene expression is "off" rather than "on," as in prokaryotes. Why is this the case? The secret lies in chromatin, or the complex of DNA and histone proteins found within the cellular nucleus. The histones are among the most evolutionarily conserved proteins known; they are vital for the well-being of eukaryotes and brook little change. When a specific gene is tightly bound with histone, that gene is "off." But how, then, do eukaryotic genes manage to escape this silencing? This is where the histone code comes into play. This code includes modifications of the histones' positively charged amino acids to create some domains in which DNA is more open and others in which it is very tightly bound up. DNA methylation is one mechanism that appears to be coordinated with histone modifications, particularly those that lead to silencing of gene expression. Small noncoding RNAs such as RNAi can also be involved in the regulatory processes that form "silent" chromatin. On the other hand, when the tails of histone molecules are acetylated at specific locations, these molecules have less interaction with DNA, thereby leaving it more open. The regulation of the opening of such domains is a hot topic in research. For instance, researchers now know that complexes of proteins called chromatin remodeling complexes use ATP to repackage DNA in more open configurations. Scientists have also determined that it is possible for cells to maintain the same histone code and DNA methylation patterns through many cell divisions. This persistence without reliance on base pairing is called epigenetics, and there is abundant evidence that epigenetic changes cause many human diseases.

For transcription to occur, the area around a prospective transcription zone needs to be unwound. This is a complex process requiring the coordination of histone modifications, transcription factor binding and other chromatin remodeling activities. Once the DNA is open, specific DNA sequences are then accessible for specific proteins to bind. Many of these proteins are activators, while others are repressors; in eukaryotes, all such proteins are often called transcription factors (TFs). Each TF has a specific DNA binding domain that recognizes a 6-10 base-pair motif in the DNA, as well as an effector domain. In the test tube, scientists can find a footprint of a TF if that protein binds to its matching motif in a piece of DNA. They can also see whether TF binding slows the migration of DNA in gel electrophoresis.

For an activating TF, the effector domain recruits RNA polymerase II, the eukaryotic mRNA-producing polymerase, to begin transcription of the corresponding gene. Some activating TFs even turn on multiple genes at once. All TFs bind at the promoters just upstream of eukaryotic genes, similar to bacterial regulatory proteins. However, they also bind at regions called enhancers, which can be oriented forward or backwards and located upstream or downstream or even in the introns of a gene, and still activate gene expression. Because many genes are coregulated, studying gene expression across the whole genome via microarrays or massively parallel sequencing allows investigators to see which groups of genes are coregulated during differentiation, cancer, and other states and processes.

Most eukaryotes also make use of small noncoding RNAs to regulate gene expression. For example, the enzyme Dicer finds double-stranded regions of RNA and cuts out short pieces that can serve in a regulatory role. Argonaute is another enzyme that is important in regulation of small noncoding RNA–dependent systems. Here we offfer an introductory article on these RNAs, but more content is needed; please contact the editors if you are interested in contributing.

Imprinting is yet another process involved in eukaryotic gene regulation; this process involves the silencing of one of the two alleles of a gene for a cell's entire life span. Imprinting affects a minority of genes, but several important growth regulators are included. For some genes, the maternal copy is always silenced, while for different genes, the paternal copy is always silenced. The epigenetic marks placed on these genes during egg or sperm formation are faithfully copied into each subsequent cell, thereby affecting these genes throughout the life of the organism.

Still another mechanism that causes some genes to be silenced for an organism's entire lifetime is X inactivation. In female mammals, for instance, one of the two copies of the X chromosome is shut off and compacted greatly. This shutoff process requires transcription, the participation of two noncoding RNAs (one of which coats the inactive X chromosome), and the participation of a DNA-binding protein called CTCF. As the possible role of regulatory noncoding RNAs in this process is investigated, more information regarding X inactivation will no doubt be discovered.

Index

A
Activation 135
Advantage 143
American 27
Amplifies 121
Analytical 44
Apparatus 109
Architecture 128
Arrangement 122
Assembly 113

B
Between 77
Bewilderingly 4
Biochemical 131

C
Changed 20
Chloroform 53
Chloroplast 97
Chromosomal 65
Chromosomes 2, 7, 58, 137
Commentaries 29
Commonly 148
Competitive 145
Complementary 37, 167
Components 133
Concentrations 56, 70, 117
Conditions 19
Configurations 188
Consequently 57
Consortium 28
Continuous 6
Creation 26

D
Degenerate 66
Depends 72
Described 51
Descriptions 91
Determined 64
Determines 71
Differences 62, 76
Different 3
Differentiate 140
Dimensional 94
Discovered 86
Discrimination 32
Diseases 1
Distinguished 73
Division 41
Dominant 13

E
Economic 34
Encourages 92
Environment 151
Environmental 132, 157
Evolutionarily 154
Experimental 179
Experiments 8, 60
Expression 102, 149, 163, 180, 185

F

Fragmontary 81
Function 139
Functional 104
Fundamental 48

G

Generation 14
Glycerophospholipids 107

H

Having 18
However 5
However 11

I

Identifying 87
Immediately 174
Important 189
Independently 116
Individually 165
Individuals 30
Information 83
Information 93
Institutions 31
Instructions 158
Intermediate 99, 100, 141
Interpretations 69
Intracellular 112, 127
Introducing 146
Investigations 36
Investigators 161

M

Maintain 142, 144
Maintenance 80
Mammalian 171
Manufacturers 54
Manufacturing 21
Measurement 147
Mechanical 119, 124

Mechanics 46
Medicine 59
Melanin 22
Membrane 114
Membrane 106, 126
Membranes 108
Mendel 49
Mentioning 12
Messenger 153, 169
Micrographs 117
Microtubule 136
Microtubules 105, 110
Microtubules 138
Modification 159, 172
Modulating 182
Molecule 164
Monastery 9
Multicellular 101
Mutations 25
Mutations 15, 17

N

Necessary 178
Neurotransmitter 38, 125, 183

P

Particular 42
Permeability 115
Phenotype 181
Phosphorylation 120
Polymerase 162, 173, 176, 177
Population 39
Possibility 134
Precipitation 55
Previous 10
Problems 16
Process 82
Prokaryotic 103
Properties 77
Proteins 78
Purple 45

Index

R
Radioactive 40
Redundant 35
Regulation 187
Regulatory 156
Relationship 95
Reproduction 150
Required 90
Responses 123

S
Segments 43
Sequence 171
Sequences 47
Sophisticated 152
Specialized 118, 130
Specifically 79, 111
Stabilization 74

Structural 89
Structure 168
Subsequent 23
Substrates 98
Synthesized 63, 84

T
Temozolomide 160
Termination 164
Transcribed 67, 184
Transcriptase 52
Transcription 175
Transcription 155, 170
Transferred 57
Transforming 61
Translation 75, 165, 166
Transmembrane 128, 129